D1118601

HIERARCHICALLY INTELLIGENT MACHINES

HIERARCHICALLY INTELLIGENT MACHINES

George N Saridis
Rensselaer Polytechnic Institute, USA

World Scientific
New Jersey • London • Singapore • Hong Kong

Published by

World Scientific Publishing Co. Pte. Ltd.

P O Box 128, Farrer Road, Singapore 912805

USA office: Suite 1B, 1060 Main Street, River Edge, NJ 07661

UK office: 57 Shelton Street, Covent Garden, London WC2H 9HE

British Library Cataloguing-in-Publication Data
A catalogue record for this book is available from the British Library.

ISBN 981-02-4790-7

Printed in Singapore by Uto-Print

To my father *Nicholas*

TABLE OF CONTENTS

Chapter 1.

INTRODUCTION

Chapter 2.

**MACHINE INTELLIGENCE, KNOWLEDGE AND PRECISION;
CONCEPTS AND DEFINITIONS**

Chapter 3.

**ENTROPY AND THE PRINCIPLE OF INCREASING PRECISION WITH
DECREASING INTELLIGENCE (IPDI)**

Chapter 4.

THE ANALYTIC FORMULATION OF HIERARCHICALLY INTELLIGENT MACHINES

Chapter 5.

HIERARCHICALLY INTELLIGENT CONTROL: THE ORGANIZATION LEVEL

Chapter 6.

HIERARCHICALLY INTELLIGENT CONTROL: THE COORDINATION LEVEL

TABLE OF FIGURES

TABLE

PREFACE

The Theory of Hierarchically Intelligent Machines originated some forty five years ago when, as a new graduate of the Athens Polytechnic, I was asked to design a "electric shaft" to raise an overflow door of a dam in Greece. The door had two motors at either end and needed to be connected for synchronization. The answer was to install two "synchroes" or synchronous three phase machines with connected rotors. This project introduced me to the concepts of Control Engineering to which I devoted the rest of my career. Intelligent Control came into the picture in 1971, when K. S. Fu coined the name for the area beyond Adaptive and Learning Control which was fashionable at that time. Its theory took shape after long discussions with Norm Caplan at our office at NSF in 1973. Intelligent Machines came later into the picture when I tried to integrate the areas of application of Intelligent Controls.

From the first years of my control experience I tried to venture into more areas more advanced in intelligence, like Adaptive and Learning Control, which were taboo in the sixties. Work in System Identification, Parameter and Performance Adaptive algorithms led into the area of Self-Organizing Control which was also the title of my first book. The idea was to collect all the methods that would provide the ground of utilizing the control experience in areas inside and outside of engineering with minimal interaction with human operators. Intelligent Control was the natural extension of this effort. Robotic systems, especially with applications to Space exploration, where it was needed, was a parallel effort in order to establish the validity of the theory. Within the limits of this work I organized the IEEE Robotics and Automation Society, the Intelligent Control Committee of the Control Systems Society, and several IFAC, Control and Robotics Conferences and Symposia. As a generalization of this effort was the extension of the theory to Intelligent Machines, with major applications to prosthetic devices, and Intelligent Manufacturing. In the meanwhile Entropy came into the picture as an integrating measure among the various engineering and non-engineering fields and an appropriate theory was developed.

I then established a pioneering, NASA supported Center for Intelligent Robotic Systems for Space Exploration, CIRSSE, at RPI, with ten professors and fifty students, and a one and a half million dollar laboratory facility. The center, in its five year existence, produced a large number of papers, technical reports and especially remote space truss assembly from the NASA Space Center at Houston, without the interference of a human operator. An Intelligent Manufacturing application for product scheduling was also performed by a student in Greece, to show the generality of the theory of Intelligent Machines.

This book summarizes the Theory of Intelligent Machines, and should be used to explore its capability to apply to engineering as well as various soft sciences, like biotechnology, ecology, pollution control and waste management, evolutionary biology, economics, etc. It is hoped that young researchers as well as professionals will pick up the flag and carry the whole Theory and Applications to new frontiers. At this point I want to thank all my students and colleagues who have contributed to the development of the material presented in this volume and especially my wife Youla whose patience and understanding made this book possible.

George N. Saridis PhD
Loudonville NY
February 2001

Chapter 1.

INTRODUCTION

1.1 MOTIVATION

In the last quarter of a century several researchers have devoted their efforts to the creation and development of machines that can perform anthropomorphic tasks in structured or unstructured environments, with little or no interaction with a human operator. The purpose of those machines was either to relieve humans from the chores of hazardous and tedious jobs or to perform tasks in remote unfamiliar and dangerous environments.

The results of such research and attempt to organize them has been recently presented in a report of a Task Force of the Control System Society of IEEE, chaired by P. Antsaklis(1993). However, most of the methods presented propose non-analytic solutions to the problem.

The **Theory of Intelligent Control systems,** the tool to implement Intelligent Machines, was developed by Saridis (1988) and combined the powerful high-level decision making of the digital computer with advanced mathematical modeling and synthesis techniques of system theory with linguistic methods of dealing with imprecise or incomplete information. This produced the desired unified approach for the design of **Intelligent Machines**. The theory may be thought of as the result of the intersection of the three major disciplines (Fig. 1.1):

- **Artificial Intelligence,**
- **Operations Research,**
- **Control Theory**.

The reason for this claim has been proven necessary is that none of the above disciplines can produce individually a satisfactory theory for the design of such machines. It is also aimed in establishing Intelligent Controls as an engineering discipline, with the purpose of designing Intelligent Autonomous Systems of the future. It combines effectively the results of cognitive systems research, with various mathematical programming control techniques.

The control intelligence is hierarchically may be distributed according to the **Principle of Precision with Decreasing Intelligence (IPDI)**, evident in all hierarchical management systems. The analytic functions of an Intelligent Machine are implemented by Intelligent Controls, using Entropy as a measure. Such an architecture is analytically implemented using entropy as a measure. However, various cost functions expressed in entropy terms, may be used to evaluate the

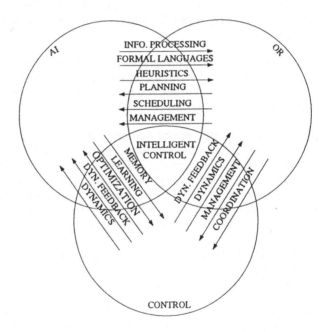

Fig. 1.1 Definition of the Intelligent Control Discipline

generated design. Reliability, a property highly desirable for systems functioning autonomously, is a very desirable measure of performance and can also be expressed by entropy, and can be combined in the criterion of performance of the system design (Saridis 2001).

This monograph introduces a learning methodology based on **the Analytic Theory of Intelligent Machines** to select one among a given set of designs to control a number of hardwired sub-systems (Saridis 1988, 1995). The above methodology is implemented by a 3-level **Hierarchical Intelligent Machine** (HIM) structured probabilistically, according to the **Principle of Increasing Precision with Decreasing Intelligence**. The Machine may use various models to implement the three hierarchically in terms of intelligence levels, which should draw from Artificial Intelligence, Operations Research, and Control fundamentals respectively. The decision probabilities at all levels are recursively updated from the **success** and **failure** signals received at the end of an execution of a command, to converge with probability 1.

The theory of Newtonian mechanics, which is the basis for the classical control theory, assumes, for fixed initial conditions, well defined deterministic motions which are reversible in time. Thus any uncertainty appearing in the system, would eventually be reduced.

On the other hand, uncertainty has been associated with insufficient knowledge and lack of information in Thermodynamics and Information Theory. The models used were probabilistic and irreversible in time and thus not deterministically reducible (Boltzmann 1876, Coveney and Highfield 1990).

Therefore, in modern physics, one may distinguish two categories of uncertainties:

*   **Subjective (reducible) uncertainties**
*   **Objective (irreducible) uncertainties**

The first category refers to the temporary lack of knowledge of the experimenter, while the latter is associated with phenomena from which no further improvement of knowledge is attainable.

Now, the control design problem will be associated to the subjective uncertainties by the following argument. The designer, is faced with the uncertainty of optimal design before tackling the problem, because he does not know yet the solution. He may assign an appropriate probability density function over the space of **admissible controls**, and reduce the design problem to the problem of finding the point (control law), that has **maximum probability of attaining the optimal value**. This approach will be pursued in the sequel with **Entropy**, as the measure of the probability of uncertainty of the design (Saridis 1988, 1995).

1.2. AUTOMATION AND CONTROL IN MODERN SOCIETY

After the Agricultural revolution of the 17th and 18th centuries, and the Industrial revolution of the 19th century, Automation is the revolution of the 20th century. It deals, in principle, with the development of machines that autonomously perform tasks that relieve the human burden of complex chores. Its primary tool is the discipline of Automatic Control, an original engineering smart process, that guides a system to perform desired tasks in a stable manner (King 1999).

The methodology of Automatic Control Systems, briefly introduced here, has been applied to engineering systems for many centuries. The Theory Automatic Control is based on the concept of feedback (Brogan 1974, James Nichols and Phillips 1947, Feld'baum 1965). From the Watt regulator to the electronic cathode coupled amplifier, control system applications have been greatly effective. However, they became popular since the Second World War when they were used to control rockets. Their success was based on the principle of Feedback, which stabilized the performance of the system. In order to make things clear to all the readers, it is appropriate to clarify certain of the concepts we are dealing: we shall start with Automation.

- **Automation is the discipline that replaces the human effort in complex tasks.**

The basis of Automation is the Computing Machine (Hilton 1963) which later, generated the Information explosion of our 20th century.

A generalized version of Automatic Control is given next, suitable to encompass recent application to disciplines outside engineering of interest in our modern society

- **Automatic Control is making a system do what you want it to do.**

Since **Feedback** is the concept central in the theory of Automatic Control, an explanation of its function is necessary

- **Feedback in Automatic Control Systems drives a system by the difference between its desired and actual output.**

The above descriptions are easily substantiated by the following analytic expressions related to Automatic Control. However, they also represent a philosophical system that finds applications in every day life. Besides the modern electronic, electrical, mechanical, manufacturing or other engineering systems they may be generalized to include ecological, environmental, economic, biological, social, administrative and other systems of general interest, by properly selecting the parameters of the analytic models of the respective system. Then, appropriate solutions may be sought through standard control methods, or others using **Entropy** as the measure of performance of the system (Brogan 1974, Baily 1990, Brooks and Wiley 1988, Faber et al. 1995, Saridis 2001).

In the 1940's, during the beginning of their applications, feedback systems, were influenced by communication technology, and were modeled in the frequency domain. Later, when state variables were introduced, they brought back the time domain models, which have the capacity of representing nonlinear, stochastic, digital, systems with discontinuities and other cases described in Chaotic systems (Prigogine 1996). Finally, optimal control was introduced in the 1960's as a mathematical tool to successfully design automatic control systems. The above ideas may be demonstrated by the following analytic presentation.

A **typical system** in an uncertain environment may be closely approximated by:

$$dx/dt = f(x,u,t) + G(t)w(t); \qquad x(0)=x_0, \qquad (1.1)$$
$$z(t) = h(x,u,t) + v(t);$$

where $x(t)$ is the state of the system at time t, $z(t)$ is the current m-dimensional measurement vector, $u(t)$ is the r-dimensional control vector, $f(x,u,t)$, $h(x,u,t)$ are twice differentiable nonlinear functions of their arguments, and $G(t)$ is an nxr matrix of known coefficients respectively. The stochastic variables $x(0)$ is the n-dimensional Gaussian initial state, $w(t)$ is the r-dimensional process noise, $v(t)$ is the m-dimensional measurement noise, defined for any $t \in [0,\infty]$, with properties:

$$E\{x(0)\} = x_0, \quad E\{(x(0) - x_0)(x(0) - x_0)^T\} = P_0, \qquad (1.2)$$
$$E\{w(t)\} = 0, \quad E\{w(t)w(\tau)^T\} = Q(t)\delta(t-\tau),$$
$$E\{v(t)\} = 0, \quad E\{v(t)v(\tau)^T\} = R(t)\delta(t-\tau).$$
$$E\{x(0)w(t)^T\} = E\{x(0)v(t)^T\} = E\{w(t)v(\tau)^T\} = 0.$$

Automatic Control which implies the finding of a control u such that the system would behave according to our instructions, may be open-loop where $u(t)$ is preprogrammed and fed as an input, or feedback where the output is compared to a desired behavior and the difference is used to drive the system.

Optimal Control may be formulated as follows. A cost functional, containing all the predefined desired specifications about the system's performance is also defined:

$$V(u) = E\{J(u)\} = E\{\varphi(x(T),T) + \int_0^T L(x,u,\tau)d\tau\} \qquad (1.3)$$

The solution of the optimal control problem is obtained by finding a control u that:

$$V(u^*) = V^* = Min_u \, I(u) \qquad\qquad (1.4)$$

Optimal control represents philosophically the attempt of human nature to obtain the best results of meeting assumed specifications.

1.3 INTELLIGENT MACHINES-A SURVEY

When one talks about Intelligent Machines it brings to his mind machines built to duplicate human-like behavior in structured environments. Examples of a human-interactive machine is the NASA Shuttle's mechanical arm (Fig. 1.2 and 1.3); and non-interactive machine SRI's "Shaky" (Fig. 1.4 and 1.5). Such machines represent an original but obsolete technology and may be thought of as the precursors of modern technology.

Modern intelligent machines are devices so labeled because they demonstrate in addition to autonomy, capability to make decisions and perform tasks with minimal human supervision. They base their operation on the function of advanced computing machines, and demonstrate some kind of Machine Intelligence(King 1999, Gupta, Singh 1996 etc.). The term Machine Intelligence means different things to different people; for instance to a large section of the literature it means the theory of Fuzzy Sets created by Lotfi Zadeh, for others it means the application of Neural Nets. It has also meant the combination into Neural-Fuzzy Systems (Lin, Lee, 1995). They all sprang out from the concepts of Knowledge-based systems and Artificial Intelligence representing a soft aspect of modern mathematics.

In contrast, a more rigorous representation of Intelligent Machines, using mathematical tools has been generated by engineering researchers, with the aid of Intelligent Control Systems (Fu, 1971, Gupta, Singh 1996 etc.). Researchers like Antsaklis, Passino, Tsypkin, Levis et. al., represented in the above reference, have

used analytic techniques encountered mostly in Automatic Control Theory have developed the theory for Machines that demonstrate "Intelligence" which may operate in unfriendly or unfamiliarly environments with minimum interaction with a human operator, thus satisfying the conditions of an Intelligent Machine.

Saridis and his colleagues have developed a theory called "Hierarchically Intelligent Control" which combines the ideas of Artificial Intelligence, and Operations Research with the ones of Automatic Control creating a compact architecture that was successfully applied to Space construction, Nuclear plant monitoring, and Manufacturing Engineering.

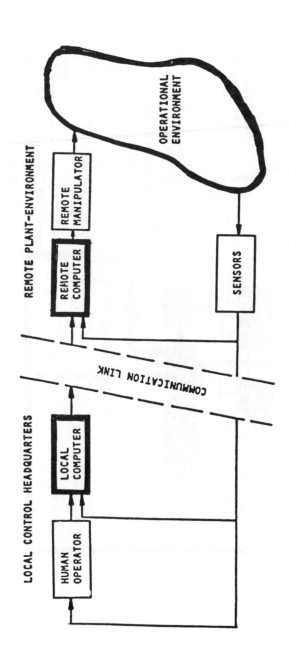

Fig. 1.2 Man-Machine Interactive Remote Controlled Manipulator

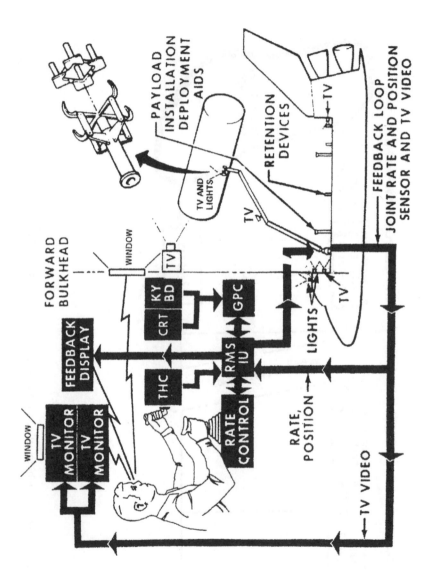

Fig. 1.3 NASA Shuttle Mechanical Arm

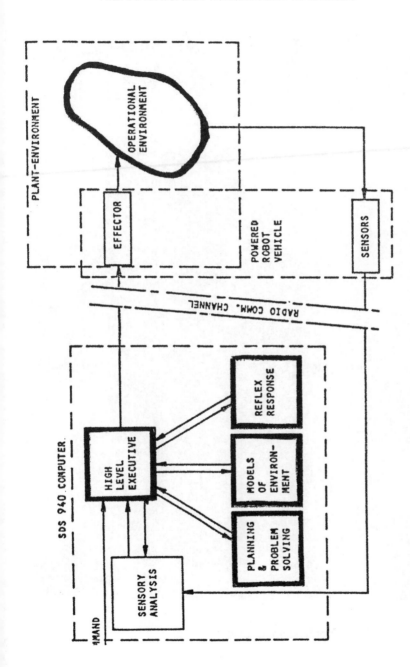

Fig. 1.4 Block Diagram of SRI's Robot System "Shaky"

Fig. 1.5 SRI's Robot System "Shaky"

The purpose of this volume is to present the methodology analytically, demonstrate its advantages and show its applications.

1.4 BOOK ORGANIZATION

This book is organized in the following manner: the present Chapter introduces to the reader, the basic concepts of Automation, Automatic Control, Intelligent Machines, and Intelligent Control Systems. It also demonstrates some philosophical aspects of the influence of the above concepts to our modern society, shown herein.

Chapter 2 describes and analyzes the essence of topics like Artificial Intelligence, Machine Intelligence, Smart Systems, Precision and Complexity and Introduces the meaning of Intelligent Control.

Chapter 3 deals with Entropy the energy function that is used as a measure of quality in the present text. It also introduces the Principle of Increasing Precision with Decreasing Intelligence a cornerstone in the development of Intelligent Machines and especially the Hierarchically Intelligent Control Systems.

Chapter 4 presents the analytic formulation of Intelligent Machines as created by the author.

Chapters 5, 6 , and 7 describe analytically the Organization, Coordination and Execution levels of the Hierarchically Intelligent Control Systems, respectively.

Chapter 8 presents the application of Hierarchically Intelligent Control Systems to Robotic Systems for Space Exploration while Chapter 9 utilizes the same concepts to Manufacturing production.

Finally Chapter 10 presents conclusions on the subject and guidance for further research. Pertinent references follow each chapter.

1.5 REMARKS

The evolution of modern science has introduced novel ideas to the interpretation of physical phenomena and integrated various previously disjoint disciplines like mechanics, thermodynamics, economics, biology, genetics, ecology, etc. (Coveney and Highfield 1990). The result was an attempt to interpret global phenomena, in a unified way, using the probabilistic theory of Automatic Control, with a common measure the underlying entropy, which permits their chaotic description in an analytic way away from their equilibrium (Prigogine 1996, Saridis 2001). As it is described in this work, this approach casts the control problem as an uncertain activity which is in agreement with the methodologies of other disciplines studied within the theory of Chaos. Time irreversibility and performance uncertainty enters the naturally the control activities. Using this philosophy, Hierarchically Intelligent Control is introduced as the way of reducing the effect of undesirable human intervention in other disciplines of science. The resulting methodology is explored in this work. It should be noticed that this presentation covers only a small area of the potential application of entropy to scientific thinking and further research should prove conclusive to the understanding of physical phenomena.

1.6 REFERENCES

Antsaklis P., et al (1993), *Final Report of the Task Force on Intelligent Control* **_IEEE Control Systems Society Magazine_** December.

Bailey K. D., (1990), **_Social Entropy Theory_**, State University of New York Press, Albany N.Y.

Boltzmann L. (1872) "Further Studies on Thermal Equilibrium between Gas Molecules" *Wien Ber.*, **66**, p. 275.

Brogan W. L.,(1974) **_Modern Control Theory_** Quantum Publishers New York N.Y.

Brooks D. R. and Wiley E. O. (1988) **_Evolution as Entropy_** University of Chicago Press, Chicago Il.

Coveney P. And Highfield R., (1990), **_The Arrow of Time_** , Fawcett Colombine, New York N.Y.

Faber M. Niemes H. Stephan G., (1995), **_Entropy, Environment and Resources_** Springer Verlag, Berlin Germany.

Feld'baum A.A. (1965), **_Optimal Control Systems_**, Academic Press, New York.

Fu K. S., (1971), "Learning Control Systems and Intelligent Control Systems: an intersection of Artificial Intelligence and Automatic Control" *IEEE Transactions on Automatic Control, **AC-16**,* (1) pp. 70-72.

Gupta, M. M. Sinha N. K., (eds), (1996), **_Intelligent Control Systems_**, IEEE Press New York NY.

Hilton A. M., (1963), **_Logic, Computing Machines, and Automation_**, Meridian Books, Cleveland and New York

James M. J., Nichols N. B., Phillips R. S., (1947), *Theory of Servomechanisms,* McGraw Hill Book Company, (MIT Radiation Laboratory), New York NY.

King, R. E., (1999), *Computational Intelligence in Automatic Control*, Marcel Dekker New York NY.

Kumar P.R., Varaiya P., (1986), *Stochastic Systems; Estimation, Identification and Adaptive Control*, Prentice Hall, Englewood Cliffs, NJ.

Lin, C-T., Lee, C. S. G., (1995), *Neural Fuzzy Systems*, Prentice Hall, Upper Saddle River NJ.

Prigogine Ilya, (1996), *La Fin des Certitudes* Editions Odile Jacob, Paris France.

Saridis G.N. (1988), "Entropy Formulation for Optimal and Adaptive Control", *IEEE Transactions on Automatic Control,* **Vol. 33**, No. 8, pp. 713-721, Aug.

Saridis G. N. (1995) *Stochastic Processes, Estimation, and Control: The Entropy Approach,* John Wiley and Sons, New York.

Saridis G. N., (2001) *Entropy in Control Engineering,* World Scientific Publishers, Singapore.

Singh, Madan G., (editor), (1987) *Systems and Control Encyclopedia; Theory, Technology Applications*, Vol. 1-8, Pergamon Press, Oxford UK.

Chapter 2.

MACHINE INTELLIGENCE, KNOWLEDGE AND PRECISION; CONCEPTS AND DEFINITIONS

2.1 INTRODUCTION

In order to properly define **Intelligent Machines** one needs to review the basic concepts.

In the last twenty or so years, a lot of discussions have taken place regarding the meaning of **Intelligence**. There are various views of the subject depending on the background of the scientist involved. The psychologists argue about **Human Intelligence**, the computer scientists talk about **Artificial Intelligence**, and the engineers stress the concept of **Machine Intelligence**. All these arguments are based on the kind of intelligence that humans demonstrate in dealing with their every day activities, a concept that is still nebulous and very little understood (Crick 1994).

If one considers **Human Intelligence** as the founding block, **Artificial Intelligence** was created to deal with the effort to make computers perform like human beings, when making decisions and perform other human like activities (Winston 1977). Finally, engineers developed the concept of **Machine Intelligence** to represent the properties of autonomous machines created to perform unsupervised anthropomorphic tasks (Saridis 1977).

The first category deals more with psychoanalytic phenomena, and it does not have to be analyzed herein even though it may serve as a guide for the other two categories. These last two though need to be well defined and categorized in order to make clear their domain of activities. This is done in the next section.

2.2 ARTIFICIAL vs. MACHINE INTELLIGENCE

The following is a paraphrase of a popular quotation (Fig. 2.1):

- **If as quoted: Artificial Intelligence is the compliment of Natural Stupidity then Machine Intelligence is the compliment of Computational Stupidity.**

There have been many arguments about the similarities and the differences of Artificial and Machine Intelligence. Artificial Intelligence being the main instrument of computer logic in performing human-like decisions in a computer environment.

FIG. 2.1 Artificial Intelligence vs. Machine Intelligence

It is also one of the three components of Machine Intelligence as defined in Chapter 1 (Fig. 1.1). However, it differs from the latter since it deals with abstract heuristic computer programming, meant to understand human behavior, while Machine Intelligence has to do with the logic of hardware designed to perform anthropomorphic tasks.

The following two definitions, due to P. H. Winston and G. N. Saridis respectively are given to clarify the subject (Winston 1977).

- **Artificial Intelligence is the study of ideas which enable computers to do the things that make people seem intelligent. Its central goals are to make computers more useful and to understand the principle, which makes Intelligence possible.**

The key components of Artificial Intelligence are: interactive systems between man and machine, heuristics and expert system exhaustive programming (Saridis 1977).

- **Machine Intelligence is the ability of a Machine to systematically organize and execute anthropomorphic tasks with the assistance of cognitive engineering systems and minimum interaction with a human operator.**

The key components of Machine Intelligence are: computer mathematics, cognitive engineering and Intelligent control.

The following two definitions, due to P. H. Winston and G. N. Saridis respectively are given to clarify the subject (Winston 1977).

- **Artificial Intelligence is the study of ideas which enable computers to do the things that make people seem intelligent. Its central goals are to make computers more useful and to understand the the principle, which makes Intelligence possible.**

The key components of Artificial Intelligence are: interactive systems between man and machine, heuristics and expert system exhaustive programming (Saridis 1977).

- **Machine Intelligence is the ability of a Machine to systematically organize and execute anthropomorphic tasks with the assistance of cognitive engineering systems and minimum interaction with a human operator.**

The key components of Machine Intelligence are: computer mathematics, cognitive engineering and Intelligent control.

The similarities and differences of Artificial and Machine Intelligence are graphically depicted in Fig. 2.2. There Artificial Intelligence is represented as a mapping of anthropomorphic tasks into the analytic tools of the computer in order to study human behavior, while Machine Intelligence is the inverse mapping of analytic tools imbedded in a machine into anthropomorphic tasks.

The similarities and differences of Artificial and Machine Intelligence are graphically contrasted in Fig. 2.2. There Artificial Intelligence is represented as a mapping of anthropomorphic tasks into the analytic tools of the computer in order to study human behavior, while Machine Intelligence is the inverse mapping of analytic tools imbedded in a machine into anthropomorphic tasks. The latter is the one that concerns the present book.

The similarities of the different functions performed by Knowledge-based Systems, Intelligent Machines and Computers are given in Table 2.1, where above the representation techniques are compared for reference

2.3 SMART SYSTEMS

Many times, various systems were designed with a capability to simply sense and modify their behavior according to the needs of their environment. The design is usually based on the ingenuity of the designer and they are preprogrammed accordingly. A typical example is the so called **Expert Systems,** which are computer programs written for specific tasks that drive the controls of a process, is a thermostat (Thomson 1985). Other such systems are the photo-sensor automatic garage door openers, the motion control house security systems, and the aircraft autopilots to mention a few.

Smart systems basically simple devices, and they reflect the Intelligence of their designer, without having the autonomy required for high level decision making, to qualify as Intelligent systems. Therefore they are not included in the present discussion. An interesting discussion of such systems is given by King (1999).

2.4 KNOWLEDGE AND INTELLIGENCE IN THE MACHINE

Some definitions regarding Machine Knowledge and Intelligence are appropriate in order to clearly define the field of Intelligent Machines. According to the **American Heritage Dictionary** *(1992)* :

• **Intelligence is defined as the capacity to acquire and apply Knowledge.**

Such a statement implies that knowledge is the key variable in an Intelligent system. Since the focus of this presentation is **Machine Intelligence,** which is

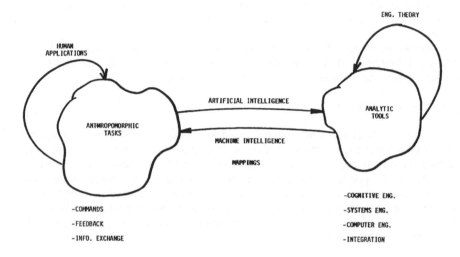

Fig. 2.2 Flows of Artificial and Machine Intelligence

	KNOWLEDGE			
KNOWLEDGE-BASED EXPERT SYSTEMS	KNOWLEDGE REPRESENTATION	REASONING	NATURAL LANGUAGES	COGNITION
INTELLIGENT MACHINES	DATA MANAGEMENT	INFERENCE	FORMAL LANGUAGES	SENSING
DIGITAL COMPUTER	MEMORY	CPU	OPERATING SYSTEM	I/O DEVICES

Table 2.1 Equivalence of Knowledge Representaiton Techniques

based on the intelligence quoted above, an appropriate definition is necessary:

- **Machine intelligence is defined as the process of analyzing, organizing and converting data into Machine Knowledge.**

Now, from reference to Shannon's Information Theory (Shannon and Weaver 1963), it is well understood that:

- **Knowledge is a form of structured Information.**

This is very convenient because an analytic formulation of Intelligent Machines may be adapted to Shannon's Information Theory. Therefore, **Machine Knowledge** is defined as follows:

- **Machine knowledge K, is defined to be structured information acquired and applied to remove ignorance or uncertainty about a specific task pertaining to an Intelligent Machine.**

Similarly,

- **The Rate of Machine Knowledge R, is the flow of Knowledge in an Intelligent Machine.**

Using the above definitions analytic expressions of Machine Knowledge and its Rate are obtained.

Assuming that Machine Knowledge is Information:

$$K = - \ln[p(K)] \tag{2.1}$$

and the average Rate of Knowledge is also:

$$R = - \alpha - \mu \ln[p(R)] \tag{2.2}$$

where $p(\cdot)$ is the probability density of the event. Solving for $p(R)$ we obtain:

$$p(R) = \exp(-\alpha - \mu R) \tag{2.3}$$

$$\alpha = \ln \int_{\Omega x} \exp(-\mu R) \, dx$$

This formulation leads ro the IPDI as it will be demonstrated in the sequel. Table 2.1 compares the various Knowledge representations.

2.5 PRECISION AND COMPLEXITY

Some additional definitions are needed to complete the foundations of **Intelligent Machines.**

Complexity is always imbedded in the design and execution of Intelligent Machines. However, their performance is always prescribed by a certain level of detail required by the task expected to be executed, which is defined as **Precision**. Such details are inversely associated with the uncertainty of execution and thus are measurable with entropy. The following definitions help to clarify this concept.

• <u>**Imprecicion**</u> **is the uncertainty of execution of the various tasks of an Intelligent Machine.**

• <u>**Precision**</u> **is the compliment of Imprecision and serves as a measure of the complexity of the process.**

The concept of precision shall be associated with the **Principle of Increasing Precision and Decreasing Intelligence (IPDI)**, a fundamental concept for Intelligent Machines, to be discussed in the next Chapter. Precision is required for thew smooth and accurate execution of tasks associated with world processes.

2.6 HIERARCHICALLY INTELLIGENT CONTROL

In order to implement the Theory of Intelligent Machines the broader definition of Automatic Control Systems presented in Chapter 1. is used.

• **Control is making a Process do what we want it to do.**

This generalization is found useful in order to accomodate nonconventional systems that are seved by Intelligent Machines, like biological, environmental etc.

• **Intelligent Control is the main tool to implement Intelligent Machines.**

Hierarchically Intelligent Controls is a special architecture defined for the design of Intelligent Machines. The Theory of Hierarchically Intelligent Controls, originally reported by Valavanis and Saridis (1992), has been recently reformulated by Saridis(1996) to incorporate new architectures that are using Neural and Petri nets Saridis (1995). The analytic functions of Hierarchically Intelligent Machines are implemented using Entropy as a measure. The resulting hierarchical control structure is based on the Principle of Increasing Precision with Decreasing Intelligence (IPDI). Each of the three levels of the Intelligent Control is using different architectures, in order to satisfy the requirements of the Principle):

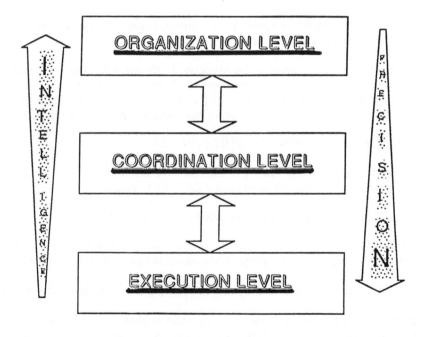

Fig. 2.3 The Structure of Intelligent Machines

The **Organization level** modeled after a Boltzmann machine for abstract reasoning, task planning and decision making;

The **Coordination level** composed of a number of Petri Net Transducers supervised by a dispatcher for command management, serving as an interface with the Organization level;

The **Execution level,** includes the sensory, navigation and control hardware which interacts one-to-one with the appropriate Coordinators, while a VME bus provides a channel for database exchange among the several devices.

This architecture is especially applicable to **Intelligent Robotic Systems** which may be described as :

• **Intelligent Robots are Intelligent Machines performing complex human-like tasks in uncertain or unfamiliar environments with minimum interaction with a human operator.**

This architecture was implemented on a robotic tele-transporter, designed for construction of trusses for the Space Station Freedom, at the Center for Intelligent Robotic Systems for Space Exploration laboratories at the Rensselaer Polytechnic Institute (Saridis 2001). The analytic formulation of Hierarchically Intelligent Controls and their applications are discussed the next chapters..

2.7 REMARKS

The basic concepts underlining the theory of Hierarchically Intelligent Machines like Machine Intelligence, Machine Knowledge, Precision and Complexity were defined and contrasted to Artificially Intelligent and Smart Systems. The basic difference being the search for an analytic formulation that would lead to an engineering implementation. Further more it has been recently realized that other scientific disciplines have being using the same concepts for an analytic representation of their subjects (Prigogine 1980). Such ideas will be discussed in the next Chapter.

2.8 REFERENCES

Hays-Roth, F., Waterman, D.A., Lenat, D.B., *Building Expert Systems*, Addison Wesley, Reading MA

King, R. E., (1999), *Computational Intelligence in Automatic Control*, Marcel Dekker New York NY.

Prigogine Ilya, (1980), *From Being to Becoming,* Freeman and Company, San Francisco.

Saridis, G. N, (1977), *Self-Organizing Control of Stochastic Systems*, Marcel Dekker New York NY.

Saridis, G. N. (1995) *Stochastic Processes, Estimation, and Control: The Entropy Approach,* John Wiley and Sons, New York.

Saridis, G. N., (1996)," Architectures for Intelligent Controls" *Chapter 6*, in *Intelligent Control Systems*, M. M. Gupta, N. K. Singh (eds) IEEE Press New York NY.

Saridis G. N., (2001) *Entropy in Control Engineering,* World Scientific Publishers, Singapore.

Shannon C., Weaver W., (1963) *The Mathematical Theory of Communications*, Illini Books, Campaign Ill.

Thomson B. And W. (1985), *Micro Expert,* IBM PC Version 1.0, McGraw Hill, New York NY.

Valavanis K. P., Saridis G. N., (1992),*Intelligent Robotic Systems: Theory, Design and Applications*, Kluwer Academic Publishers, Boston MA.

Winston P. H. (1984), *Artificial Intelligence,* Addison Wesley, Reading MA.

Chapter 3.

ENTROPY AND THE PRINCIPLE OF INCREASING PRECISION WITH DECREASING INTELLIGENCE (IPDI)

3.1 UNCERTAINTY AND ENTROPY

Entropy is a form of lower quality energy, first encountered in Thermodynamics. It represents an undesirable form of energy that is accumulated when any type of work is generated. Recently it served as a model of different types of energy based resources, like transmission of information, biological growth, environmental waste, etc. Entropy was currently introduced, as a unifying measure of performance of the different levels of an Intelligent Machine by Saridis (2001). Such a machine is aimed at the creation of **modern intelligent robots which may perform human tasks with minimum interaction with a human operator**. Since the activities of such a machine are energy related, entropy may easily serve as a cost measure of performing various tasks as Intelligent Control, Image Processing, Task Planning and Organization, and System Communication among diversified disciplines with different performance criteria. The model to be used is borrowed from Information Theory, where the uncertainty of design is measured by a probability density function over the appropriate space, generated by an equivalent of Jaynes' Maximum Entropy Principle.

Other applications of the Entropy concept are the definition of Reliability measures for design purposes and measures of complexity of the performance of a system, useful in the development of the theory of Intelligent Machines.

Due to its wide applicability, entropy has been used as a convenient global measure of performance to a large variety of systems of diverse disciplines including waste processing, environmental, socio-economic, biological and other. Thus, by serving as a common measure, it may expand system integration by incorporating say societal, economic or even environmental systems to engineering processes (Baily 1990, Brooks and Wiley 1988, Faber et al. 1995, Prigogine 1990, Rifkin 1989, Saridis 2001)..

In this Chapter, Entropy concepts will be mainly used for the reformulation of the theory of stochastic optimal control and its approximation theory, state estimation, and parameter identification used in Adaptive Control systems. It will prove Fel'dbaum's claim that the stochastic Open-Loop-Feedback-Optimal and Adaptive Control, based on the **Certainty Equivalence Principle** (Kumar & Varaiya 1986), (Morse 1990, and 1992), are not optimal. It will also provide a measure of goodness of their approximation to the optimal in the form of an upper bound of the Entropy missed by the approximation.

An application to the design of Intelligent Machines using performance measures based on Reliability and Complexity will demonstrate the power of the approach.

Uncertainty, thus far, has been associated with insufficient knowledge and lack of information in Thermodynamics and Information Theory. The models used were probabilistic. As in Chapter 1, one may distinguish two categories of uncertainties:

*   **Subjective (reducible) uncertainties**
*   **Objective (irreducible) uncertainties**

The first category refers to the lack of knowledge of the experimenter, while the latter refers to phenomena from which no further improvement of knowledge is attainable.

Now, the control design problem will be associated to subjective uncertainty. The designer, is faced with the uncertainty of optimal design before solving the problem. He may assign an appropriate probability density function over the space of **admissible controls**, and reduce the design problem into the problem of finding the point (control law), that has **maximum probability of attaining the optimal value**. This approach will be pursued in the sequel with **Entropy**, the measure of uncertainty.

Uncertainties have also been expressed in terms of possibilistic models, like:

*   **Fuzzy Sets** (Zadeh 1994)
*   **Mathematical Theory of Evidence** (Stephanou 1986)

However, such forms are not encounter in this work and will not be considered.

Other attempts have also been made to formulate the estimation problem (Kalata, Premier 1974), and the feedback control problem (Wiedemann 1969), using entropy methods. However, those attempts found little success in the literature.

3.2 ENTROPY AND THERMODYNAMICS: BOLTZMANN

The concept of Entropy was introduced in Thermodynamics by Clausius in 1867, as the low quality energy resulting from the second law of Thermodynamics. This is the kind of energy which is generated as the result of any thermal activity, at the lower thermal level, and is not utilized by the process.

It was in 1872, though, that Boltzmann used this concept to create his **theory of Statistical Thermodynamics**, thus expressing the uncertainty of the state of the molecules of a perfect gas. The idea was created by the inability of the dynamic theory to account for all the collisions of the molecules, which generate the thermal energy. Boltzmann (1872) stated that the entropy of a perfect gas, changing states isothermally, at temperature T is given by;

$$S = - k \int_x (\psi-H)/kT \exp\{(\psi-H)/kT\} \, dx \qquad (3.1)$$

where ψ is the Gibbs energy, $\psi = - kT \ln \exp \{-H/kT\}$, H is the total energy of the system, and k is Boltzmann's universal constant. Due to the size of the problem and the uncertainties involved in describing its dynamic behavior, a probabilistic model was assumed where the Entropy is a measure of the molecular distribution. If p(x) is defined as the probability of a molecule being in state x, thus assuming that,

$$p(x) = \exp\{(\psi-H)/kT\} \qquad (3.2)$$

where p(x) must satisfy the "incompressibility" property over time, of the probabilities, in the state space X, e.g.;

$$dp/dt = 0 \qquad (3.3)$$

The incompressibility property is a differential constraint when the states are defined in a continuum, which in the case of perfect gases yields the Liouville equation. Substituting eq.(3.2) into eq.(3.1) the Entropy of the system takes the form,

$$S = - k \int_x p(x) \ln p(x) \, dx \qquad (3.4)$$

The above equation defines Entropy as a measure of the uncertainty about the state of the system, expressed by the probability density exponential function of the associated energy.

Actually, the problem of describing the entropy of an isothermal process should be derived from the Dynamical Theory of Thermodynamics, considering heat as the result of the kinetic and potential energies of molecular motion. It is the analogy of the two formulations that led into the study of the equivalence of entropy with the performance measure of a control system. If the Dynamical Theory of Thermodynamics is applied on the aggregate of the molecules of a perfect gas, an Average Lagrangian I, should be defined to describe the average performance over time of the state x of the gas,

$$V = \int_0^{tf} L(x,t) \, dt \qquad (3.5)$$

where the Lagrangian L(x,t) = (Kinetic energy) - (Potential energy). The Average Lagrangian when minimized, satisfies the **Second Law of Thermodynamics**.

Since the formulations eqs.(3.1) and (3.5) are equivalent, the following relation should be true;

$$S = V/T \qquad (3.6)$$

where T is the constant temperature of the isothermal process of a perfect gas (Lindsay and Margenau, 1957). This relation will be the key in order to express the performance measure of the control problem as Entropy.

3.3 ENTROPY AND INFORMATION THEORY: SHANNON

In the 1940's Shannon (1963), using Boltzmann's idea, e.g., eq. (3.4), defined Entropy (negative) as a measure of the uncertainty of the transmission of information, in his celebrated work on **Information Theory**:

$$H = - \int_{\Omega} p(s) \, \ln p(s) \, ds \qquad (3.7)$$

where $p(s)$ is a Gaussian density function over the space Ω of the information signals transmitted. The similarity of the two formulations is obvious, where the uncertainty about the state of the system is expressed by an exponential density function of the energy involved.

Shannon's theory was generalized for dynamic systems by Ashby (1965), Boettcher and Levis (1983), and Conant (1976) who also introduced various laws which cover information systems, like the Partition Law of Information rates.

3.4 KOLMOGOROV'S ε-ENTROPY

The ε-entropy formulation of the metric theory of complexity, originated by Kolmogorov (1956) and applied to system theory by Zames (1979) is another use of entropy.

It implies that an increase in knowledge about a system, decreases the amount of ε-entropy which measures the uncertainty (complexity) involved with the system.

$$\varepsilon - H = \ln(n_e) \qquad (3.8)$$

where n_e is the minimum number of coverings of a set e. Therefore ε-entropy is a measure of complexity of the system involved. It may also be interpreted as a measure of precision if it viewed as the number of points required to describe a line as in Fig. 3.1.

3.5 ENTROPY, ENVIRONMENT AND MANUFACTURING

Since the latest major improvements in the average quality of life, major increases have occurred in the production of waste, traffic congestion, biological pollution and in general environmental decay(Rifkin 1980), which can be interpreted as the increase of the Global Entropy of our planet, an energy that tends to deteriorate the quality of our modern society. According to the second axiom of thermodynamics this is an irreversible phenomenon, and nothing can be done to eliminate it.

This section introduces optimal control, developed for systems engineering, to environmental and manufacturing systems to effectively restrain the growth of the Global Entropy. Since the paper is addressed to the nonspecialist reader, an attempt will be made to introduce the concepts of systems, automatic control, optimal control and global entropy for information purposes. Then a formal presentation will be made, of the proposed theory developed from an entropy point of view which will relate the optimal control theory to the Global Entropy, and thus present a method to minimize its effect to our society. This theory has in addition to the practical applications, a philosophical foundation that has implications to the quality of life and the future of our planet.

3.6 THE MODIFIED JAYNES' PRINCIPLE OF MAXIMUM ENTROPY

In an attempt to generalize the principle, used by Boltzmann and Shannon, to describe the uncertainty of the performance of a system under a certain operating condition, Jaynes (1957) formulated his **Maximum Entropy Principle**, to apply it in Theoretical Mechanics. In summary it claims that

•   **The uncertainty of an unspecified relation of the function of a system is expressed by an exponential density function of a known energy relation associated with the system.**

A modified version of the Principle, as it applies to the Control problem, is derived in the sequel, using Calculus of Variations (Saridis 1987). The proposed derivation represents a new formulation of the control problem, either for deterministic or stochastic systems and for optimal or non-optimal solutions.

The purpose of this work is to establish entropy measures, equivalent to the performance criteria of the optimal control problem, while providing a physical meaning to the latter. This is done by expressing the problem of control system design probabilistically and assigning a distribution function representing the uncertainty of selection of the optimal solution over the space of admissible controls. By selecting the worst case distribution, satisfying **Jaynes' Maximum Entropy Principle**, the performance criterion of the control is associated with the entropy of selecting a certain control (Jaynes 1957), (Saridis 1985). Minimization of the differential entropy, which is equivalent to the average performance of the system, yields the optimal control solution. Furthermore, the Generalized Hamilton-Jacobi-Bellman equation is derived from the incompressibility over time condition of the probability distribution. Adaptive control and stochastic optimal control are obtained as special cases of the optimal formulation, with the differential entropy of active transmission of information, claimed by Fel'dbaum (1965), as their difference. Upper bounds of the latter may yield measures of goodness of the various stochastic and adaptive control algorithms. In this section, the entropy measure for optimal control will be established.

The optimal feedback deterministic control problem with accessible states is defined as follows: given the dynamic system;

$$dx/dt = f(x,u,t) \; ; \; x(t_0) = x_0; \qquad\qquad (3.8)$$

and the cost function,

$$V(u;x_0,t_0) = \int_{t0}^{T} L(x,u,t) \, dt \qquad\qquad (3.9)$$

where $x(t)\varepsilon\Omega_x$ is the n-dimensional state vector $u(x,t)\varepsilon\Omega_u XT \subset \Omega_x XT$, is the m-dimensional feedback control law and $t \; \varepsilon \; \mathfrak{I} = [t_0, T]$.

An optimal control $u^{\cdot}(x,t)$ is sought to minimize the cost,

$$V(u^{\cdot};x_0,t_0) = \text{Min}_u \int_{t0}^{T} L(x,u,t) \, dt \qquad\qquad (3.10)$$

Define the differential entropy, for some $u(x,t)$,

$$H(x_0,u(x,t),p(u)) = H(u) = -\int_{\Omega x0}\int_{\Omega x} p(x_0,u)\ln p(x_0,u) \, dudx_0 \qquad\qquad (3.11)$$

where $x_0\varepsilon\Omega_{x0}$, $x\varepsilon\Omega_x$ the spaces of initial conditions and states respectively, and $p(x_0,u)=p(u)$ the probability density of selecting u. One may select the density function $p(u)$ to maximize the differential entropy according to **Jaynes' Maximum Entropy Principle** (Jaynes 1957), subject to $E\{V(x_0,u,t)\}=K$, for some $u(x,t)$. This represents a problem more general than the optimal where K is a fixed but unknown constant, depending on the selection of $u(x,t)$. The unconstrained expression of the differential entropy is;

$$I = \beta H(u) - \gamma[E\{V\}-K] - \alpha\int_{\Omega x} p(u)dx-1\Big] = -\int_{\Omega x} [\beta p(u)\ln p(u)+\gamma p(u)V]dx - \alpha\Big[\int_{\Omega x} p(u)dx -1\Big] \qquad (3.12)$$

Using the Lemma of Calculus of Variations, maximization of I with respect to $p(u)$ yields,

$\partial/\partial[-\beta p(u)\ln p(u) - \gamma p(u)V - \alpha p(u)] = 0; \quad \partial^2 I/\partial p^2 < 0$

or

$-\beta\ln p(u) - \beta - \gamma V - \alpha = 0; \qquad\qquad -\beta/p(u) < 0$

and the worst case density is,

$p(u) = e^{-\lambda - \mu V(u(x,t),x0,t0)}$

$$e^\lambda = \int_{\Omega x} e^{-\mu V(u(x,t),x0,t0)}\, dx \qquad\qquad (3.13)$$

with $\mu = \gamma/\beta$, $\lambda = (\alpha+1)/\beta$, and

$$E\{V(u(x,t),x_0,t_0)\} = K = \partial[\ln\int_{\Omega x} e^{-\mu V}\, dx]/\partial\mu$$

Now let us recast the optimal feedback control problem as the "optimal design" problem under the worst uncertainty of selecting the best control, from the set of admissible feedback controls, $u(x,t) \varepsilon \Omega_u \times T \subset \Omega_x \times T$. It is assumed that the set of admissible controls is covered by the probability density function p(u), given by eq. (3.13), expressing the uncertainty of selection of the optimal function. The worst case (maximum) differential entropy corresponding to p(u) is given by,

$$H(u) = \lambda + \mu E\{V(u(x,t),x_0,t_0)\} \qquad\qquad (3.14)$$

and the corresponding minimum value with respect to u(x,t) represents the optimal design. This condition implies,

$$dH/du = \partial H/\partial u + \partial H/\partial p\ \partial p/\partial u = 0,\ \text{(Frechet Derivative)} \qquad\qquad (3.15)$$

However, Jaynes' Principle implies,

$$\partial H/\partial p = 0.$$

Therefore,

$$\bar{H}\,[\,\dot{\phi}\,/\,u\,] = C + E\,[\,\jmath\,\frac{\partial V}{\partial \phi}\,(\,\dot{\phi}-\phi\,) + (\,\dot{\phi}-\phi\,)^T\,[\,0.5\,\jmath\,\frac{\partial^2 V}{\partial \phi^2} + Q\,]\,(\,\dot{\phi}-\phi\,)\,\}H\,[\,\dot{\phi}\,/\,u\,]$$

$$dH/du = \partial H/\partial u \qquad\qquad (3.16)$$

3.7 THE PRINCIPLE OF INCREASING PRECISION DECREASING INTELLIGENCE

After the discussion of all the above concepts we may now state the Principle of Increasing Precision with Decreasing Intelligence (IPDI). In most organization systems, the control intelligence is hierarchically distributed from the highest level which represents the manager to the lowest level which represents the worker. On the other hand, the precision or skill of execution is distributed in an inverse manner from the bottom to the top as required for the most efficient performance of such complex systems. This was analytically formulated as the **Principle of Increasing Precision with Decreasing Intelligence (IPDI)**, by Saridis (1988). The formulation and proof of the principle is based on the concept of Entropy.

For MI the Machine Intelligence, DB a data base, and R the Rate of Machine Knowledge, (Saridis 1989):

$$Prob(MI,DB) = Prob(R)$$

Then,

$$Prob(MI/DB) \times Prob(DB) = Prob(R)$$

$$\ln Prob(MI/DB) \times \ln Prob(DB) = \ln Prob(R)$$

Taking the expected values of both sides, and considering $H(\cdot)$ to be the associated entropy:

$$H(MI/DB) + H(DB) = H(R) \qquad (3.17)$$

And if MI is independent of the data base DB, then:

$$H(MI) + H(DB) = H(R) \qquad (3.18)$$

The above procedure represents the derivation of the Principle of Increasing Precision with Decreasing Intelligence, which suggests that an intelligent system requires less intelligence when it retains high precision (complexity), in order to produce the same flow of knowledge. Formally this is expressed by the statement:

• **Machine Intelligence (MI) is the set of actions and/or rules which operates on a Data-base (DB) of events or activities to produce flow of knowledge.**

$$(MI) : (DB) \Rightarrow (R)$$

This principle defines the structure of the Hierarchically Intelligent Machines, and will be explored in the sequel.

3.8 REMARKS

Entropy is defined as a Universal Energy which results from the production of Work in a system. The production of Entropy is irreversible without the use of additional work, and may represent thermal energy in Thermodynamics, Information in Communication systems, Performance in Control systems, as well as waste and pollution in Ecological systems, Economic spending in Societal systems, or Biodegradation in Biological systems (Boltzmann 1872, Shannon 1963, Rifkin 1980, Prigogine 1980, Saridis 1985, Brooks 1989, Bailey1990, Farber et al 1990). It represents an unifying measure for globalization of of many disjoint sciences. It may successfully be used as measure for the development of Hirarchically Intelligent Machines, using the **Principle of Increasing Precision with Decreasing Intelligence** as the structural model.

3.9 REFERENCES

Ashby W. R., (1975), *An Introduction to Cybernetics*,J. Wiley & Sons, Science Edition, New York.

Bailey K. D., (1990), *Social Entropy Theory*, State University of New York Press, Albany N.Y.

Boettcher K. L., Levis A. H., (1983),"Modeling the Interacting Decision-Maker with Bounded Rationality", *IEEE Transactions on System Man and Cybernetics*, *Vol. SMC-12*, 3, pp. .

Boltzmann L. (1872) "Further Studies on Thermal Equilibrium between Gas Molecules" *Wien Ber.,* **66**, p. 275.

Brooks D. R. and Wiley E. O. (1988) *Evolution as Entropy* University of Chicago Press, Chicago Il.

Faber M. Niemes H. Stephan G., (1995), *Entropy, Environment and Resources* Springer Verlag, Berlin Germany.

Conant, R. C., (1976), "Laws of Information which Govern Systems", *IEEE Transactions on System Man and Cybernetics*, *Vol. SMC-6*, No. 4, pp. 240-255.

Feld'baum, A.A. (1965), *Optimal Control Systems*, Academic Press, New York.

Jaynes, E.T. (1957), "Information Theory and Statistical Mechanics", *Physical Review, Vol.4*, pp. 106.

Kalata, P. Premier, R., (1974), "On Minimal Error Entropy Stochastic Approximation" *International Journal of System Science, Vol. 5*, No. 9, pp. 985-986.

Kumar, P.R., Varaiya, P., (1986), *Stochastic Systems; Estimation, Identification and Adaptive Control,* Prentice Hall, Englewood Cliffs, NJ.

Lindsay, R.B., Margenau, (1957), *Foundations of Physics*, Dover Publications, New York NY.

Prigogine, I., (1980), *From Being to Becoming*, W. H. Freeman and Co. San Francisco, CA.

Kolmogorov, A.N. (1956), "On Some Asymptotic Characteristics of Completely Bounded Metric Systems", *Dokl Akad Nauk*, SSSR, *Vol. 108*, No. 3, pp. 385-389.

McInroy J.E., Saridis G.N.,(1991), "Reliability Based Control and Sensing Design for Intelligent Machines", in *Reliability Analysis* ed. J.H. Graham, Elsevier North Holland, N.Y.

Morse, A.S., (1990), "Towards a Unified Theory of Parameter Adaptive Control: Tunability", *IEEE Trans. on Automatic Control, Vol. 35*, No. 9, pp. 1002-1012, September.

Morse, A.S., (1992), "Towards a Unified Theory of Parameter Adaptive Control-Part II: Certainty Equivalence and Implicit Tuning", *IEEE Trans. on Automatic Control, Vol. 37*, No. 1, pp. 15-29, January.

Rifkin J., (1989), *Entropy into the Greenhouse World*, Bantam Books, New York NY.

Saridis, G.N. (1979), "Toward the Realization of Intelligent Controls", *IEEE Proceedings, Vol. 67*, No. 8.

Saridis, G. N. (1983), "Intelligent Robotic Control", *IEEE Trans. on Automatic Control, Vol. 28*, No. 4, pp. 547-557, April.

Saridis, G.N. (1985), "An Integrated Theory of Intelligent Machines by Expressing the Control Performance as an Entropy", *Control Theory and Advanced Technology, Vol. 1*, No. 2, pp. 125-138, Aug.

Saridis, G.N. (1988), "Entropy Formulation for Optimal and Adaptive Control", *IEEE Transactions on Automatic Control*, *Vol. 33,* No. 8, pp. 713-721, Aug.

Saridis, G.N. (1989), "Analytic Formulation of the IPDI for Intelligent Machines", *AUTOMATICA the IFAC Journal, 25*, No. 3, pp. 461-467.

Saridis, G.N., (1996)," Architectures for Intelligent Controls" *Chapter 6*, in *Intelligent Control Systems*, M. M. Gupta, N. K. Singh (eds) IEEE Press New York NY.

Saridis G. N., (2001) *Entropy in Control Engineering,* World Scientific Publishers, Singapore.

Saridis, G.N. and Graham, J.H. (1984), "Linguistic Decision Schemata for Intelligent Robots", *AUTOMATICA the IFAC Journal, 20*, No. 1, pp. 121-126, Jan.

Saridis, G.N. and Lee, C.S.G. (1979), "Approximation of Optimal Control for Trainable Manipulators", *IEEE Trans. on Systems Man and Cybernetics*, *Vol.8*, No. 3, pp. 152-159, March.

Shannon, C. and Weaver, W. (1963), *The Mathematical Theory of Communications*, Illini Books.

Stephanou H. E., (1986), "Knowledge based control systems" *IEEE Workshop on Intelligent Cotrol*, RPI Troy NY, p116

Tsai, Y.A., Casiello, F.A., Loparo, K.A., (1992), "Discrete-time Entropy Formulation of Optimal and Adaptive Control Problems", *IEEE Transactions on Automatic Control, Vol. 37*, No. 7, pp. 1083-1088, July.

Valavanis, K.P.,Saridis, G.N., (1992), *Intelligent Robotic Systems: Theory and Applications,* Kluwer Academic Publishers, Boston MA.

Wang, F., Saridis, G.N. (1990) "A Coordination Theory for Intelligent Machines" *AUTOMATICA the IFAC Journal, 35,* No. 5, pp. 833-844,Sept.

Weidemann, H.L., (1969), "Entropy Analysis of Feedback Control Systems" in *Advances in Contol Systems*, C. Leondes Ed., Academic Press.

Zames, G. (1979), "On the Metric Complexity of Casual Linear Systems, ϵ-entropy and ϵ-dimension for Continuous Time", *IEEE Trans. Automatic Control*, *24*, No. 2, pp. 220-230, April.

Zadeh L. A.,(1994), "Fuzzy Logic, Neural Networks, and Soft Computing"
Communications of the ACM 37(3) , PP. 77-84

Chapter 4.

THE ANALYTIC FORMULATION OF HIERARCHICALLY INTELLIGENT MACHINES

4.1 INTRODUCTION

In the last few years **Hierarchically Intelligent Machines,** proposed by Saridis (1977), have reached a point of maturity to be implemented on a robotic test bed aimed for space assembly and satellite maintenance as well as nuclear power plant surveillance and scheduling in manufacturing. They feature an application of the **Theory of Hierarchically Intelligent Control,** based on the **Principle of Increasing Precision with Decreasing Intelligence (IPDI)** and form an analytic methodology, using Entropy as a measure of performance. The original structural architecture is a three level system, designed according to the above principle, and the software architecture is based on an information theoretic approach (Saridis 1996).

Many researchers have being proposing different approaches to design Intelligent Machines, based on single disciplines of intelligence, like optimal control, automata theory or artificial intelligence. However, optimal control is not well equipped to perform analytically, complex decision functions and automata theory and artificial intelligence cannot perform precise manipulations of hardware or other human tasks. Until something like that will happen, it is necessary to find solutions to the Intelligent Machine problem by combining the above disciplines.

Thus Saridis has proposed the following postulate for the theory of Hierarchically Intelligent Control, in order to design Intelligent Machines:

- The theory of Hierarchically Intelligent Control is the *mathematical problem* of finding the right sequence of decisions and controls for a system structured according to the *Principle of Increasing Precision with Decreasing Intelligence* such that its *total Entropy* is Minimal.

Minimization of the total Entropy should be performed by repetitive reevaluation of the cost (entropy) after a three level execution of a given command. Such an optimization procedure, using Hierarchical Stochastic Automata Theory, is discussed in detail by Lima and Saridis (1996). However, such a process is tedious and time consuming. Instead, observing that the time scales of each level different, e.g. one complete iteration of one level is one step of the previous one, a simplification is proposed: to minimize each level separately. This approximation affects very little the overall optimization, and produces an efficient procedure for the solution of the problem.

4.2 THE ARCHITECTURE OF THE MACHINE

A tri-discipline architecture of the Hierarchically Intelligent Machine is composed of three levels, resembling the efficient structure of the human brain and motor system:

- **Organization level**
- **Coordination level and**
- **Execution level**

It represents the original architecture of the system (Valavanis and Saridis 1992), with
minor modifications in their internal structures, recently implemented to incorporate more efficient and effective developments, dictated by experience (Fig. 4.1).

This volume discusses these new architectures for each one of the levels separately, and justifies their effectiveness by presenting some implementation results from the robotic transporter in CIRSSE at RPI (Saridis 1995), and in scheduling in manufacturing systems (Varvatsoulakis et al 2000).

4.3 DEFINITIONS OF THE PROCEDURE

The proposed architecture is modeled after the Theory of Organization Systems (Boetcher, Levis 1983). With the help of the **Principle of Increasing Precision with Decreasing Intelligence (IPDI)**, it involves a streamlined pyramidal structure where the manager sits on the top with endowed intelligence and few manual skills, for abstract decision making. The workers are at the other end of the pyramid with a lot of skills and little intelligence needed. In between "foremen" serve as interface between the two (Fig.4.1).

Another interesting feature of this architecture is the **task composition from primitive events,** instead of the typical task decomposition of other methods, constituting a **bottom up approach.**

This architecture is claimed to be much more efficient than others that have appeared in the literature (Albus 1975, Antsaklis 1994, Meystel 1986) because of the efficient (pyramidal) structure of the various levels. It covers by far more general applications than the typical robotic ones, like scheduling, management etc. Following the above theory, only three generic levels have been defined. They are precisely defined as follows:

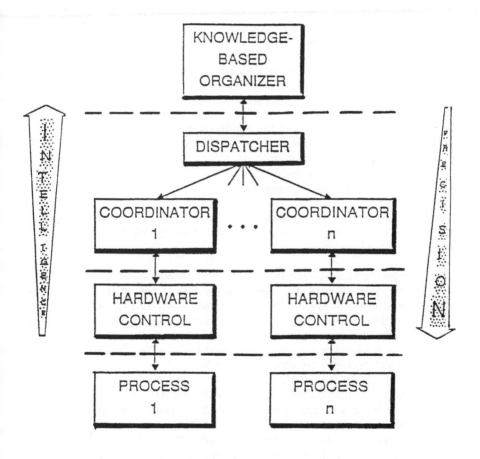

Fig. 4.1 A Hierarchically Intelligent Control System

The Organization Level is the engine where abstract planning and decision making is performed. Its structure is based on concepts from Artificial Intelligence and performs the following specific functions, using the Flow of Knowledge as the system's variable measured by the associated Entropy;

This architecture is claimed to be much more efficient than others that have appeared in the literature (Albus 1975, Antsaklis 1994, Meystel 1986) because of the efficient (pyramidal) structure of the various levels. It covers by far more general applications than the typical robotic ones, like scheduling, management etc. Following the above theory, only three generic levels have been defined. They are precisely defined as follows:

The Organization Level's structure is based on concepts from Artificial Intelligence and performs the following specific functions, using the Flow of Knowledge as the system's variable measured by the associated Entropy;

- **Machine Reasoning and Knowledge Representation**
- **Machine Task Planning**
- **Machine Decision Making**
- **Long-term Memory Control**
- **Machine Learning (Feedback)**

The word "Machine" is used to distinguish these functions from the ones defined by Artificial Intelligence. It generates a series of commands for each plan to be executed by the lower levels.

The Coordination Level Performs the functions of control and communication of commands, generated by the Organization Level, among a reconfigurable **Dispatcher** and the fixed structure **Coordinators** which correspond to each of the hardware of the lower level. These Coordinators select the most efficient algorithms for execution of the appropriate plan. Communication with the environment is also performed at this level through the lower level hardware. The functions performed by the Coordination level are summarized as follows:

- **Reconfigurable Dispatching and Task Interpretation.**
- **Fixed Structure Coordination for each individual Hardware Device**
- **Tree Structure Command and Control Communication**
- **Reliability and Learning through Feedback**
- **Remote Dispatching with no Data appearing at this Level**

Feedback through the lower level is performed to effectively update the level's parameters.

The **Execution level** is composed of the hardware necessary for the execution of the task assigned by the Organization level and involves the respective algorithms selected by the Coordinators. It utilizes Control system, sensing or any other theoretically developed schemes for managing the hardware, using Entropy as a measure. Feedback provided through this level is appropriately translated and fed back to the Coordination level for adjustments. The main functions of the Execution level, generated for a robotic system are:

- **Vision system**
- **Motion Control**
- **Sensory Hardware**
- **Grasping system**
- **Data Communication Bus**
- **Individual Hardwired and Software Packages for Local Data Processing.**

The Maximum Entropy formulation for Optimal Control, to be discussed in Chapter 7, was specifically developed for this purpose.

4.4 HIERARCHICAL MODELING OF THE LEVELS OF THE INTELLIGENT MACHINE

The structures that may be used in the various levels are fixed but arbitrary, subject to the specifications stated in motivation section of Chapter 1, so that a designer may apply his or her preferred software and hardware. This is demonstrated by the two versions of the architecture developed by the author and his colleagues in Valavanis and Saridis (1992) and Saridis (1996). Other schemes are also acceptable as long as they conform with the basic rules of the Theory of Intelligent Machines. Actually, at the Haifa meeting in 1992, it was agreed by Albus, Antsaklis, Harmon, Meystel, and Saridis, all designers of Intelligent Machines, that their algorithms may be thought of as legitimate alternate designs of Hierarchically Intelligent Machines (Antsaklis 1994).

In this volume the most recent architecture designed by Saridis (1994), will be presented. However, the Hierarchical model of Valavanis and Saridis (1992) will also be mentioned in the sequel.

The Organization level represents an abstract design of tasks to be executed from a given command, and is based on functions drawn from Artificial Intelligence, like Machine Planning and Representation Decision Making etc. It may be the same for similar tasks executable by the other level for appropriate world configurations the strings operations are usually transmitted to a **Dispatcher** the next level.

The Coordination level is composed of a number of **Coordinators,** driven by the Dispatcher and corresponding to the devices of the next level, communicated semantically, as shown in Fig. 4.2. The time frame of the Coordination level is much faster than the Organization level. Since for one of its commands corresponds a whole search and algorithm selection of the Coordination level. Therefore, the Organization level is quasi-static for the Coordination level and can be optimized independently.

The Execution level contains the necessary hardware and the appropriate algorithms to drive them. This way data trafficking is avoided among the levels avoiding communication jams. Again the time frame of the Execution level is much faster than that of the Coordination level permitting independent optimization. Feedback of the completed tasks is used to upgrade the previous levels.

4.5 REMARKS

A very efficient design of Hierarchically Intelligent Machines, proposed by Saridis, was outlined in this Chapter. It, is general enough to accept various specific algorithms according to the taste of each designer, who follows the IPDI and the tri-discipline architecture. The next Chapters will present the most recent design, using a Boltzmann Machine for the organization level, Petri net Transducers for the Coordination level, and Entropy formulated optimal control and sensing for the Execution level, which was proposed by Saridis (1996), in order to implement the Theory of Hierarchically Intelligent Machines and its applications.

4.6 REFERENCES

Albus, J.S. (1975), "A New Approach to Manipulation Control: The Cerebellar *ransactions of ASME, J. Dynamics Systems, Measurement and Control, 97,* pp. 220-227.

Antsaklis P. Chair (1994), "Defining Intelligent Control" Report of the Task Force on Intelligent Control, *IEEE Control Systems Magazine* Vol. 14, No. 3, p. 4.

Boettcher K. L., Levis A. H., (1983),"Modeling the Interacting Decision-Maker with Bounded Rationality", *IEEE Transactions on System Man and Cybernetics, Vol. SMC-12*, 3, pp. .

Lima P. U., Saridis G. N., (1996), *Design of Intelligent Control Systems Based on Hierarchical Stochastic Automata*, World Scientific Publishers, Singapore.

Meystel, A. (1986), "Cognitive Controller for Autonomous Systems", *IEEE Workshop on Intelligent Control 1985*, p. 222, RPI, Troy, New York.

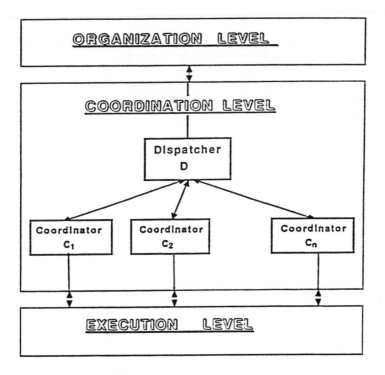

Fig. 4.2 Topology of the Coordination Level

Saridis, G.N. (1977), *Self-Organizing Controls of Stochastic Systems*, Marcel Dekker, New York, New York.

Saridis G. N. (1995) *Stochastic Processes, Estimation, and Control: The Entropy Approach,* John Wiley and Sons, New York.

Saridis, G.N., (1996)," Architectures for Intelligent Controls" *Chapter 6,* in *Intelligent Control Systems*, M. M. Gupta, N. K. Singh (eds) IEEE Press New York NY.

Valavanis, K.P., Saridis, G.N., (1992), *Intelligent Robotic Systems: Theory and Applications,* Kluwer Academic Publishers, Boston MA.

M. N. Varvatsoulakis, G. N. Saridis, P. N. Paraskevopoulos, (2000), "Intelligent Organization for Flexible Manufacturing" *IEEE Transactions on Robotics and Automation* Vol.16, No. 2, pp. 180-189, April.

Chapter 5.

HIERARCHICALLY INTELLIGENT CONTROL: THE ORGANIZATION LEVEL

5.1 INTRODUCTION

The **Organization level**, is the highest level of the architecture of Hierarchically Intelligent Machines. It represents the master control planner of the system, and interprets in an abstract fashion, a given command for action.

There are several ways to implement the Organization level of a Hierarchically Intelligent Control System, provided that the Principle of Increasing Precision with Decreasing Intelligence (IPDI) is followed. A **Knowledge-based Expert system** with a rule controlled inference engine could be a suitable choice for such a design. An alternative approach is the Probabilistic model, proposed by Valavanis and Saridis (1992), where the Rule-based architecture is replaced by a probabilistic structure governing the same functions. Another possible formulation of the Organization level is described in the **Information Theoretic modeling of Intelligent Robots by** Valavanis and Saridis in a paper of (1988)(Fig. 5.1).

However, a Boltzmann type Neural net was found recently (Saridis 1996), to be more efficient in performing the functions of the Organization level. It can perform all the required tasks by that level, and transmit them efficiently to the coordination level. Its analytic description is given in the sequel.

5.2 THE BOLTZMANN NEURAL NET

A **Boltzmann machine** type neural net, originally proposed for text generation, has been used for the structure that implements the Organization level of an Intelligent Machine developed by Saridis and Moed(1988,1992). This machine would connect a **finite** number of letters (nodes) into grammatically correct words (rules), by minimizing at the first layer the total entropy of connections. Replacing the letters at the nodes with words, at second layer, sentences are created. At the third level the words are replaced by sentences at the nodes and so on and so forth until a meaningful text is created.

The functions of the Organizer, following the model of a knowledge based system, comprise of **representation, abstract task planning** (with minimal knowledge of the current environment), **decision making, and learning** from experience. All those functions can be generated by a Boltzmann machine similar to the text generating machine, by considering a finite number of primitive elements at the nodes, constituting the basic actions and actors at the **representation phase**.

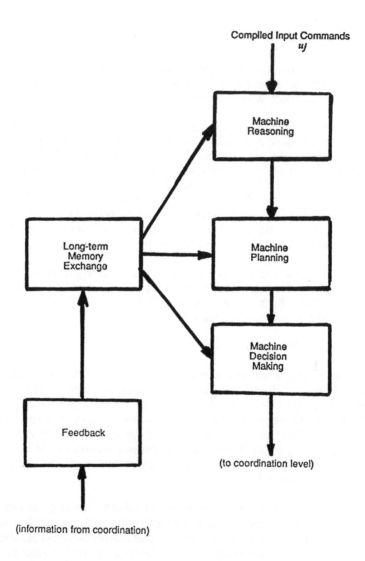

Fig. 5.1 Typical Architecture of the Organization Level

Strings of these primitives are generated by the Boltzmann machine at the **planning phase** with the total entropy representing the cost of connections. The selection of the string with minimum entropy is the **decision making** process, and the upgrading of the parameters of the system by rewarding the successful outcomes through feedback, is the **learning** procedure. The next to minimum entropy string may be retained as an alternate plan in case of failure of the original or errors created by the environment.

This bottom-up approach, characteristic of natural languages, is extremely simple and effective, utilizing intelligence to replace the complexity of the top-down type task decompositions. The tasks thus generated, are practically independent of the current environment. Information about the present world should be gathered at the Coordination level. An appropriate world model is constructed from sensory and motion information available at that level. However, there the structure of the Dispatcher, designed to interpret the Organizer's strings, monitor and traffic commands among the other Coordinators is highly dependent on the strings which represent the planned tasks.

5.3 THE ANALYTIC MODEL

To specify analytically the model of the organizer, it is essential to derive the domain of the operation of the machine for a particular class of problems as in Saridis and Valavanis (1988). Assuming that the objects of the environment are known, one may define the following functions on the organization level:

a. Machine Representation and Abstract Reasoning, (RR) is the association of the compiled command to a number of activities and/or rules. A probability function is assigned to each activity and/or rule and the Entropy associated with it is calculated. When rules are included one has active reasoning (inference engine).

In order to generate the required analytic model of this function the following sets are defined:

The set of **commands** $C = \{c_1, c_2, ..., c_q\}$ in natural language, is received by the machine as inputs. Each command is compiled to yield an equivalent machine code explained in the next section.

The **task domain** of the machine contains a number n of independent objects.

The set $E = \{e_1, e_2, ..., e_m\}$ are **individual primitive events** stored in the long-term memory and representing primitive tasks to be executed. The task domain indicates the capabilities of the machine.

The set $A = \{a_1, a_2, ..., a_l\}$ are **individual abstract actions** associating the above events to create sentences by concatenation. They are also stored in the long-term memory.

The set $S = \{s_1, s_2, ..., s_n\} = E \cup A$, $n=m+l$, is the group of **total objects** which combined, define actions represent complex tasks. They represent the nodes of a Neural net.

A set of **random variables** $X = \{x_1, ..., x_n\}$ representing the state of events is associated with each individual object s_i. If the random variable x_i is binary (either 0 or 1), it indicates whether an object s_i is **inactive** or **active**, in a particular activity and for a particular command. If the random variables x_i are continuous (or discrete but not binary) over [0,1], they reflect a membership function in a fuzzy decision making problem. In this work, the x_i's are considered to be binary.

A set of **probabilities** P associated with the random variables X is defined as follows:

$$P = \{P_i = \text{Prob}[x_i = 1]; \; I=1,...n\} \tag{5.1}$$

The probabilities P are known at the beginning of the representation stage. In order to reduce the problem of dimensionality a subset of objects is defined for a given command c_k:

$$S_k = \{s_i; \; P_i \geq a: i=1...n\} \subset S \tag{5.2}$$

b. Machine Planning,(P), is ordering of the activities. The ordering is obtained by properly concatenating the appropriate abstract primitive objects $s_i \in S_k$ for the particular command c_k, in order to form the right abstract activities (sentences or text).

The ordering is generated by a Boltzmann machine, which measures the average flow of knowledge from node j to node i on the Neural-net by

$$R_{ij} = -\alpha_{ij} - \tfrac{1}{2}E\{w_{ij}x_ix_j\} = -\alpha_{ij} - \tfrac{1}{2}w_{ij}P_iP_j \geq 0 \tag{5.3}$$

The probability due to the uncertainty of knowledge flow into node i, is calculated as in Saridis (1996):

$$p(R_i) = \exp(-\alpha_i - \tfrac{1}{2} \sum_j w_{ij}P_iP_j) \tag{5.4}$$

where

$w_{ij} \geq 0$ is the interconnection weight between nodes i and j
$w_{ij} = 0$
$\alpha_i > 0$ is a probability normalizing factor.

The average Flow of Knowledge Ri into node i, is:

$$R_i = \alpha_i + \tfrac{1}{2}E\{ \sum_j w_{ij}x_ix_j\} = \alpha_i + \tfrac{1}{2} \sum_j w_{ij}P_iP_j$$

with probability $P(R_i)$, (Jaynes' Principle 1954):

$$P(R_i) = \exp[-\alpha_i - \tfrac{1}{2} \sum_j w_{ij}P_iP_j]$$

The Entropy of Knowledge Flow in the machine is

$$H(R) = - \sum_j [P(R_i) \ln[P(R_i)] = \sum_j [\alpha_i + \tfrac{1}{2} (\sum_j w_{ij}P_iP_j) \exp[-\alpha_i - \tfrac{1}{2} \sum_j w_{ij}P_iP_j] \quad (5.5)$$

The normalizing factor α_i is such that $\tfrac{1}{2}^n \leq P(R_i) \leq 1$.

The entropy is maximum when the associated probabilities are equal, $P(R_i i) = \tfrac{1}{2}^n$ with n the number of nodes of the network. By bounding $P(R_i)$ from below by $\tfrac{1}{2}^n$ one may obtain a unique minimization of the entropy corresponding to the most like sequence of events to be selected.

Unlike the regular Boltzmann machines, this formulation does not remove α_i when $P_i = 0$. Instead, the machine operates from a base entropy level $\alpha_i\exp(-\alpha_i)$ defined as the **Threshold Node Entropy** which it tries to reduce (Saridis,Moed 1988).

c. Machine Decision Making,(DM) is the function of selecting the sequence with the largest probability of success. This is accomplished through a search to connect a node ahead that will minimize the Entropy of Knowledge Flow at that node:

$$H(R_i) = (\alpha_i + \tfrac{1}{2} (\sum_j w_{ij}P_iP_j) \exp[-\alpha_i - \tfrac{1}{2} \sum_j w_{ij}P_iP_j]$$

A modified genetic algorithm, involving a global random search, has been proposed by Moed and Saridis (1992), as a means of generating the best sequence of events that minimized the uncertainty of connections of the network expressed by the entropy (5.5).This algorithm, proven to converge globally compared favorably with other algorithms like the Simulated Annealing and the Random Search.

d. Machine Learning,(ML) (Feedback). Machine Learning is obtained   by feedback devices that upgrade the probabilities P_i and the weights w_{ij} by evaluating the performance of the lower levels after a successful iteration.

For y_k representing either P_{ij} or w_{ij}, corresponding to the command c_k, the upgrading algorithms are:

$$y_k(t_k+1) = y_k(t_k) + \beta_k(t_k+1)[\Gamma(t_k+1) - y_k(t_k)] \qquad (5.6)$$

$$J_k(t_k+1) = J_k(t_k) + \sigma_k(t_k+1)[V^k_{obs}(t_k+1) - J_k(t_k)]$$

where $J_k(t_k)$ is the performance estimate, V^k_{obs} is the observed value and

$$P_i : \Gamma_k(t_k+1) = x(t_k)$$

$$w_{ij} : \Gamma_k(t_k+1) = \begin{cases} 1 & \text{if } J = \min J_e \\ e & \\ 0 & \text{otherwise} \end{cases} \qquad (5.7)$$

e. Memory Exchange (ME), is the retrieval and storage of information from the **long-term memory**, based on selected feedback data from the lower levels after the completion of the complex task.

The above functions may be implemented by a two level Neural net, of which the nodes of the upper level represent the primitive objects s_i and the lower level of primitive actions relating the objects of a certain task. The purpose of the organizer may be realized by a search in the Neural net to connect objects and actions in the most likely sequence for an executable task.

Since it was agreed to use Petri Net Transducers (PNT) to model the coordinators at the next level, a Petri Net generator is required to create the Dispatcher's PNT for every task planned. This can be accomplished by another Boltzmann machine or a part of the existing plan generating architecture.

A graph of the Boltzmann machine with the appropriate symbols is given in Fig. 5.2. The Optimum path of the Boltzmann Machine is sought by a search technique. A list of such search techniques is given in the sequel. The **Modified Genetic Algorithm** was selected as most suitable for the present Organization level.

5.4 SEARCH TECHNIQUES

There are various search techniques that may be used to produce an optimal task generation in the Organization level of an Intelligent Machine. Three of those

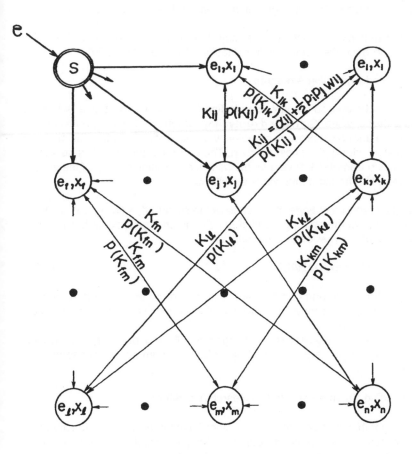

e_i = primitive event

x_i = state of event e_i, $c\{1, 0\}$; with prob. p_i

K_i = energy at node i, = $\alpha_i + \dfrac{1}{2} \sum\limits_{j} p_i p_j w_{ij}$

w_{ij} = learned weights

$p(K_{ij})$ = probability of connection i-j

Fig. 5.2 The Boltzmann Machine for the Organization Level

techniques, adapted to yield global minima of the associated Entropy, are presented here. By searching among the visible neurons in the minimum entropy state of the Boltzmann Machine, one may determine the sequence of primitive events that produce the string of optimal tasks (Moed, Saridis 1990).

5.4.1 Simulated Annealing Algorithm.

The search technique commonly used to find the minimum cost of a Boltzmann Machine Is the so called "**Simulated Annealing**" algorithm. The method is heuristic simulating the annealing process of metal by probabilistically finding the minimum value of a state- dependent cost function. The control of the search randomness depends on a parameter T driving for the minimum, "ground state", of the entropy of the system H. The algorithm works as follows:

Given a small random change of the state of the system $X_i = \{x_1, x_2,...x_n\}$ to X'_i the resulting change of the entropy is ΔH. If $\Delta H \leq 0$ the change is accepted; if $\Delta H > 0$ the probability of the new state is defined as

$$p(X_{i+1} = X'_i) = \exp(-\Delta H/K_n T) \tag{5.8}$$

where K_n is the Boltzmann constant and T is the set by the user parameter. By reducing T along an "annealing" schedule the system should settle to a near ground state as $T \to 0$.
The method being purely heuristic converges usually to a local minimum, without a rigorous proof of convergence and to a global minimum only under some strict conditions.

5.4.2 Expanding Subinterval Random Search Algorithm

A search algorithm that converges to a global minimum of a cost function of a dynamic system, is the "Random Search Technique" (Saridis 1977). Using again the entropy H, as the cost function for a given state X_i, one may define the following random search algorithm with an appropriately selected parameter μ:

$$X_{i+1} = \begin{cases} X'_i & \text{if} \quad H(X'_i) - H(X_i) \leq 2\mu \\ X_i & \text{if} \quad H(X'_i) - H(X_i) > 2\mu \end{cases} \tag{5.9}$$

where H(X) is the entropy introduced by the state $X = \{x_1, x_2,...x_n\}$, and X'_i is a randomly selected state vector generated by a prespecified independent and identically distributed density function.

It was shown that (Saridis 1977):

$$\lim_{n \to \infty} \text{Prob } \{H(X_n) - H^*_{min} < \delta\} = 1 \qquad (5.10)$$

where H^*_{min} is the global minimum entropy of the system. The existence of the minimum has been proven in (Saridis 1977).

5.4.3 Genetic Search Algorithm

Holland's Genetic Algorithm (Holland 1975), is developed by representing each point in the state space with a changing population, by a binary string and has an associated cost, dictated by the system cost function at that point to be minimized. The notation used is:

W	=	Population of members (points),
W'	=	New population of members,
|W|	=	Number of members in W,
W_k	=	kth member of the population W,
$W_k(m)$	=	mth bit of W_k,
J_k	=	cost of W_k,
S_k	=	Probability of member k being selected from current population,
J_{max}	=	Maximum cost of a possible member in W,
n	=	Length of W_k in bits.

Each iteration of a Basic Genetic Algorithm proceeds as follows:

1) Compute J_k for all $W_k \in W$,
2) Let $J_k' = J_{max} - J_k$, $\forall k$. Compute $S_k = J_k' / (\Sigma_l J_k')$, $\forall k$,
3) Repeat until $|W'| = |W|$,
    a) Randomly select W_j, W_k from W, based on S_j, S_k.
    b) Randomly generate an index $l \in \{1...n\}$
        Exchange the right string halves of W_j, W_k, called "crossover" or"matting"
        (i.e. $W_j'(i..n) = W_k(i..n)$ and $W_k'(l..n) = W_j(i..n)$)
    c) Place W_j', W_k' in W. Return W_j, W_k to W.
4) Set W = W'.
5) Repeat until $W_k \in W$ has a minimum cost.

In an attempt to prevent premature population convergence a **mutation operator** is introduced to the system. With a new generation of the population, each bit of every member has a small probability of inverting. he inversion adds diversity to the population and promotes search in previous unexplored regions of the space in an attempt to find the global minimum.

The algorithm is based on the idea of a fixed small population W. If W has very large or infinite members with many local minima the algorithm may be stuck to a part of the space and yield only a local minimum cost. They also tend to converge prematurely for that set of functions. To correct these a modification is proposed by the next algorithm which is proven to converge to the global minimum.

5.4.4 Modified Global Genetic Algorithm

In order to avoid the difficulties arising in the **Genetic Search Algorithm** and prove global convergence of the method, Moed and Saridis (1990) proposed the **Modified Genetic Algorithm**. In addition or instead of the **mutation operator** an **immigration operator** is used. A member $W_i \in W'$ is introduced, generated by the **random search algorithm** discussed in section 5.4.2 from a uniform density function, every M iterations for some $M > 0$.

If $W_k \in W$ and for all $W_i \in W$, $J_k \leq J_i$, then $W_k \in W'$. In other words, the best performing (minimum cost) string in the current population is placed in the next population. The Algorithm develops as follows:

1) The **Expanding Subinterval random search algorithm** is embedded into the **Genetic Algorithm**

2) The function $C(x) = \{y: Z(y) \leq Z(x)\}$ of the **Genetic Algorithm** is guaranteed.

3) The combined **Expanding Subinterval random search algorithm and Genetic Algorithm** yields the **Modified Genetic Algorithm** which according to theorem by Luenberger (1984) converges in probability since the **Expanding Subinterval random search algorithm** does so.

The above conditions do not bind the new algorithm severely. The **immigration rate** (immigrations/iteration), 1/M, is related to the **mutation rate** (mutations/bit) as follows:

$$M = (\text{mutations/bit}) * (\text{members/population}) \qquad (5.11)$$

and the immigration of new members may be generated with probability 1/M. By changing the **immigration rate**, the algorithm adjusts from global to local search. Even though this rate is fixed for intervals of the search, high **immigration rate** will force the algorithm towards the **random search**, while low **immigration rate** will force it towards the **Genetic Algorithm**. This algorithm was proven successful in most situations.

5.4.5 A Test Case for Comparisons

A net of 15 connected nodes was created to test the previously discussed methods (Moed Saridis 1990). Entropy was minimized in the ensuing simulations to compare these algorithms. Nodes 4 and 6 formed the input to the network. The values of these nodes were held at 1, corresponding to a desired goal input. By changing the values of the other nodes the minimum entropy of the net work could be found invoking the various search techniques. At minimum energy, the value of the 13 output nodes formed an ordered binary string, which is the correct set of primitive events of an **Organization level** for a given goal input of value 1. For this case the network had three entropy minima, corresponding to the states:

(001010100100100, 110110110001101, 001111101100010)

which where the strings of primitive events. The respective entropy of these states were (0.8, 0.6, 1.0). The global entropy minimum 0.6 was sought by all the search techniques.

In the **Simulated Annealing** algorithm the cooling temperature at time t $T_i(t)$ was used, with T_0 the initial temperature:

$$T_i(t)/T_0 = 1/\log(10+t)$$

The **Expanding Subinterval random search algorithm** used the following probability function $P(x_i)$ at node x_i, with initial value 0.5:

$$P(x_i=1)_{t+1} = \begin{cases} P(x_i=1)_t + 0.1^*[1.0 - P(x_i=1)_t] & \text{For active node} \\ P(x_i=1)_t + 0.1^*P(x_i=1)] & \text{For inactive node} \end{cases}$$

The **Modified Genetic Algorithm** was performed with population set at 20 members, each 15 bits long, so that each population had 300 bits. The immigration rate was set to 0.5 which yielded a mutation rate of 0.025.

The simulations presented here show the best and worst performance over ten trials. The results are depicted in the Figs. 5.3, 5.4, 5.5, 5.6, 5.7, 5.8.

The **Modified Genetic Algorithm** found the minimum entropy string between the 20th and 180th population, after generating 400 and 2000 points respectively. The best performance of **Simulated Annealing** required 5500 generated points to find the minimum while its worst performance did not find it after 12000 points. The best performance of the **Random Search** found the minimum after 2000 generated points, while its worst performance did not find it after 12000 points. This result indicates the superiority of the **Modified Genetic Algorithm**.

5.5 REMARKS

This Chapter presented the analytic formulation of the Organization level of an

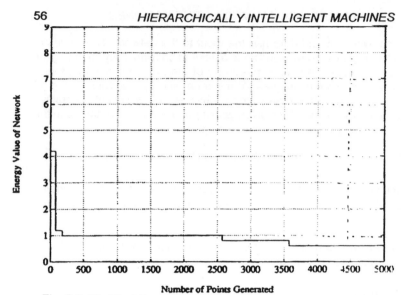

Fig. 5.3 Modified Genetic Algorithm: The Worst case

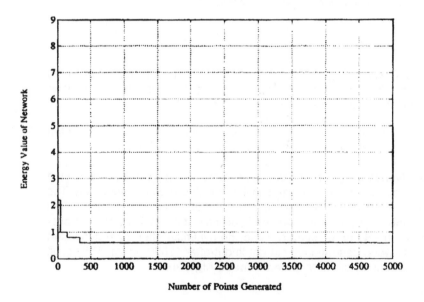

Fig. 5.4 Modified Genetic Algorithm: The Best Case

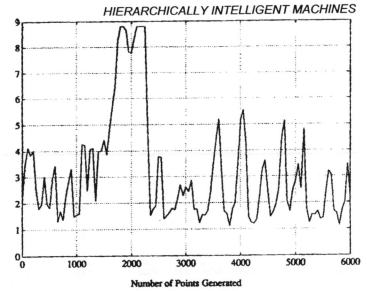

Fig. 5.5 Simulated Annealing: The Worst case

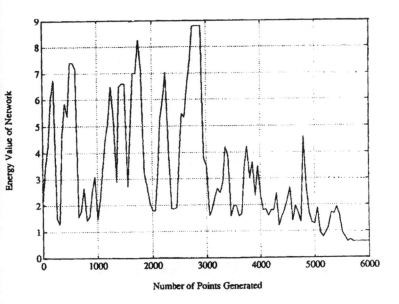

Fig. 5.6 Simulated Annealing: The Best case

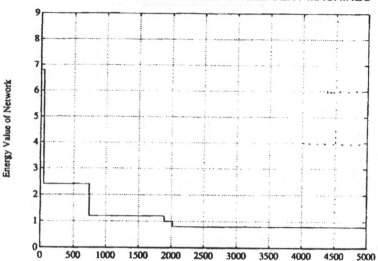

Fig. 5.7 Expanding Subinterval Random Search Algorithm: The Worst case

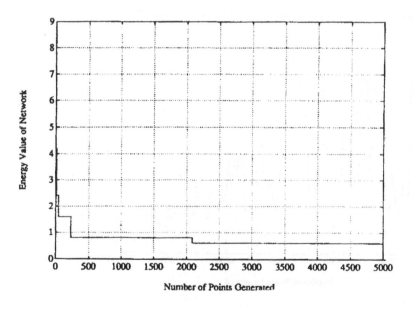

Fig. 5.8 Expanding Subinterval Random Search Algorithm: The Best case

Intelligent Machine. Even though various methods like Expert Systems, Information Theory etc. may be used to implement such a device which has as a goal to generate abstract plans for the performance of the Intelligent Machine, emphasis was given to a new Boltzmann Machine with nodes representing abstract primitive events with associated random states and te probabilities of their inclusion. Transition weights were also include in the network. Various search techniques were discussed in the sequel to search for the string of nodes that minimize the total entropy of the network. However, local search at each node to find the transition with minimum entropy from one node to another has also been used.

Another form of a Boltzmann machine with nodes confined in vertical planes, representing the various alternatives at each workstation is explored in Chapter 9 of the present book. Such a machine is suitable for production scheduling in an industrial plant and represents a variation of the present configuration.

5.6 REFERENCES

Antsaklis P. Chair (1994), "Defining Intelligent Control" Report of the Task Force on Intelligent Control, *IEEE Control Systems Magazine* Vol. 14, No. 3, p. 4.

Holland J. H., (1975), *Adaptation in Natural and Artificial Systems* The University of Michigan Press, Ann Arbor MI.

Jaynes, E.T. (1957), "Information Theory and Statistical Mechanics", *Physical Review*, pp. 106, 4.

Luenberger D. L., (1984), *Linear and Nonlinear Programming, Second Edition*, Addison Wesley, Reading MA.

Moed, M.C. and Saridis, G.N. (1990), "A Boltzmann Machine for the Organization of Intelligent Machines", *IEEE Transactions on Systems Man and Cybernetics, 20*, No. 5, Sept.

Saridis G. N. (1977), "Expanding Subinterval Random Search for System Identification and Control" *IEEE Transactions on Automatic Control*, pp.405-412.

Saridis, G.N., (1996)," Architectures for Intelligent Controls" *Chapter 6*, in *Intelligent Control Systems*, M. M. Gupta, N. K. Singh (eds) IEEE Press New York NY.

Saridis, G.N. and Moed, M.C. (1988), "Analytic Formulation of Intelligent Machines as Neural Nets", *Symposium on Intelligent Control*, Washington, D.C., August.

Valavanis, K.P., Saridis, G.N., (1988), "Information Theoretic modeling of Intelligent Robotic Systems" *IEEE Transactions on Systems Man and Cybernetics*, Vol. 18, *No.6*.
Valavanis, K.P., Saridis, G.N., (1992), ***Intelligent Robotic Systems: Theory and Applications***, Kluwer Academic Publishers, Boston MA.

Chapter 6.

HIERARCHICALLY INTELLIGENT CONTROL: THE COORDINATION LEVEL

6.1 INTRODUCTION

The next level of the Hierarchically Intelligent Machine is the **Coordination level**. It serves as an intermediary between the Organization and the Execution levels, by interpreting the generated string of abstract commands to real world tasks to be assigned to the Execution level. It is organized according to the hardware devices available, and carries the addresses only of the execution programs, leaving those software packets to the next level. This will prove helpful for remote operations.

6.2 THE ARCHITECTURE OF COORDINATION

The **Coordination level** is a tree structure of **Petri Net Transducers** as coordinators, proposed by Wang and Saridis(1988)with the Dispatcher as the root. Fig. 6.1 depicts such a structure. The Petri Net Transducer for the Dispatcher is generated by the Organizer for every specific plan and is transmitted, asynchronously, to the Coordination level along with the plan to be executed. The function of the Dispatcher is to interpret the plan and assign individual tasks to the other coordinators, monitor their operation, and transmit messages and commands from one coordinator to another as needed. As an example, a command is sent to the vision and sensing coordinator to generate a model of the environment, the coordinates of the objects for manipulation to be tabulated, and then transmitted to the motion coordinator for navigation and motion control. This command is executed by having each transition of the associated Petri Nets to initialize a package corresponding to a specific action. These packages are stored in short memories associated with each of the coordinators.

The rest of the coordinators have, in many cases, a fixed structure with alternate menus available at request. (Varvatsoulakis et al1998) Variable structure coordinators may be used if this is necessary, as in manufacturing planning (Varvatsoulakis et al 2000). They communicate commands and messages with each other, through the Dispatcher. They also provide information about reception of a message, data memory location, and job completion.

No data is communicated at the Coordination level, s.ince the task planning and monitoring may be located in a remote station, and such an exchange may cause a channel congestion. A preferred configuration for such situations is that the coordinators with a local dispatcher may be located with the hardware at the work site, while a remote dispatcher, connected to the organizer, interacts with local one

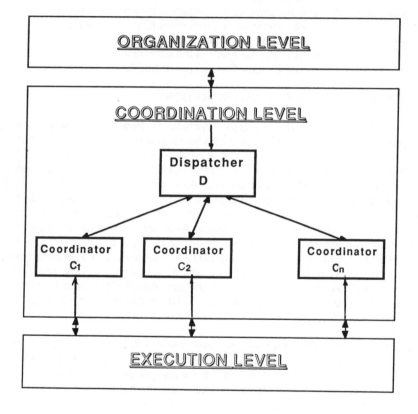

Fig. 6.1 The Architecture of the Coordination Level

from a remote position. Fig. 6.2 depicts this architecture. This concept simplifies the communication problem considerably, since only short messages are transmitted back and forth through a major channel between local and remote stations, requiring a narrow bandwidth. An example of the effectiveness of such an architecture may be demonstrated in space construction, where robots work in space while task planning and monitoring is done on earth.

Even though, there is no limitation to the number of coordinators attached to the Dispatcher, only the following ones are planned for an Intelligent Robot for space applications.

<u>Vision and Sensory Coordinator.</u> This device coordinates all the sensory activities of the robot, with cameras and lazers, and generates information of the world model in Cartesian coordinates.

<u>Motion Control Coordinator.</u> This device receives control, object and obstacle information and uses it to navigate and move multiple robotic arms and other devices, for object manipulation and task execution. It also assigns the appropriate operations on the data acquired for the desired application.

<u>Planning Coordinator.</u> The task plans, optimal and alternate gene-rated by the Organizer are stored in this device for proper monitoring of execution and possible error recovery in cases of failure of the system.

<u>Grasping Coordinator.</u> his device coordinates the grippers of the arms and interfaces the proximity sensors for effective grasping.

Entropy measures, are developed by McInroy and Saridis(1991), at each coordinator such that they may be used to minimize the complexity and improve the reliability of the system. A typical PNT system for the Coordination level of an Intelligent Robot as proposed by Wang and Saridis (1988), is given in Fig. 6.3.

6.3 PETRI NETS AND PETRI NET TRANSDUCERS

Petri nets have been proposed as devices to communicate and control complex heterogenous processes. These nets provide a communication protocol among stations of the process as well as the control sequence for each one of them. Abstract task plans, suitable for many environments are generated at the organization level by a grammar created by Wang and Saridis (1990):

$$G = (N, \textstyle\sum_o, P, S) \tag{6.1}$$

where

The upper level plan   The feedback to the upper level

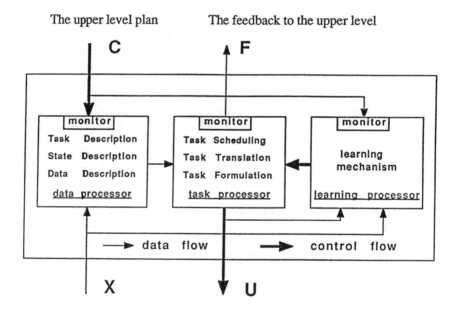

Fig. 6.2 Uniform Design of the Dispatcher and Coordinators

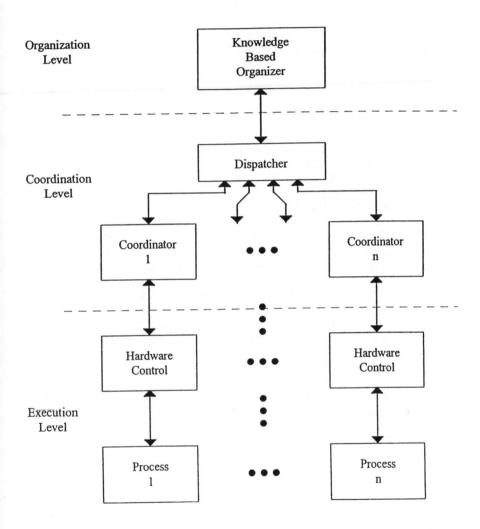

Fig. 6.3 Architecture of Intelligent Coordination

N = {S, M, Q, H} = Non-terminal symbols
$\sum_o$ = {A_1, A_2,...A_n} = Terminal Symbols (activities)
P = Production rules

Petri Net Transducers (PNT) proposed first by Wang and Saridis (1988), are Petri net realizations of the **Linguistic Decision Schemata** introduced by Saridis and Graham (1984), as linguistic decision making and sequencing devices. They are defined as 6-tuples:

$$M = (N, \sum, \Delta, \sigma, \mu, F) \qquad (6.2)$$

where

N = (P, T, I, O) = A Petri net with initial marking μ,
$\sum$ = a finite input alphabet
Δ = a finite output alphabet
σ = a translation mapping from T x ($\sum$ U {λ}) to finite sets of Δ^*; and
F C R(μ) a set of final markings.

A **Petri Net Transducer (PNT)** is depicted in Fig. 6.4. Its input and output languages are **Petri Net Languages (PNL)**. In addition to its on-line decision making capability PNT's have the potential of generating communication protocols, learning by feedback, ideal for the communication and control of coordinators and their dispatcher in real time. Petri Net Transducers use Entropy as a measure, calculated according to Fig. 6.5 and discussed in detail in Saridis(2001).A typical architecture of a coordination level is given in Fig. 6.6, and may follow a scenario suitable for the implementation of an autonomous intelligent robot.

6.4 THE COORDINATION STRUCTURE

Fig. 6.7 depicts the Petri Net Structure of a typical **Coordination Structure (CS)** of an intelligent robot. This structure is a 7-tuple:
......
$$CS = (D, C, F, R_D, S_D, R_C, S_C) \qquad (6.3)$$

where

D = (N_d, $\sum_o$, Δ_o, G_d, μ_d, F_d) = The PNT dispatcher
C = {C_1,...C_n} = The set of coordinators
C_i= (N^i_c, $\sum^i_c$, Δ^i_c, σ^i_c, μ^i_c, F^i_c) = the ith PNT coordinator
F = $U^n_{i=1}${f_I, f_{SI}, f_O, f_{SO}} = A set of connection points
R_D,R_C = Receiving maps for dispatcher and coordinators
S_D,S_C = Sending maps for dispatcher and coordinators

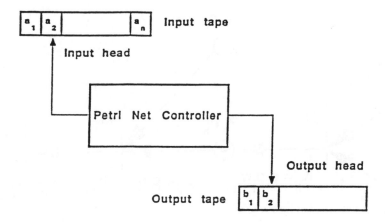

Fig. 6.4 A Petri Net Transducer (PNT)

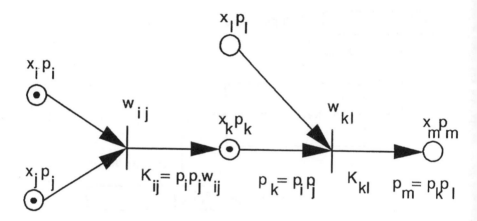

Fig. 6.5 Entropy Measures of a Petri Net Transducer

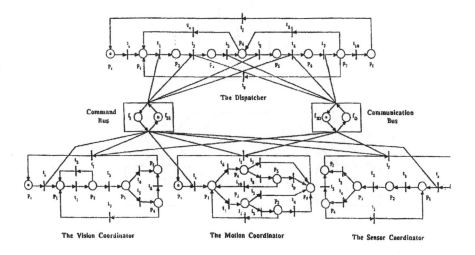

Fig. 6.6 Petri Net Diagram of a Typical Coordination Level

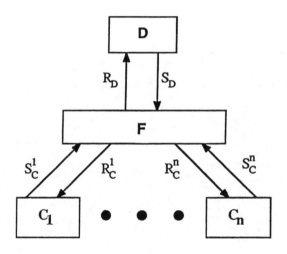

Fig. 6.7 A Typical Coordination Structure

6.5 TASK SCHEDULING AND TRANSLATION

Decision making in the coordination structure is accomplished by **Task Scheduling** and **Task Translation**, e.g., for a given task find σ an enabled t such that σ(t,a), is defined and then select the right translation string from σ(t,a) for the transition t.

The sequence of events transmitted from the organization level is received by the dispatcher which requests a world model with coordinates from a vision coordinator. The vision coordinator generates appropriate database and upon the dispatcher's command communicates it to the planning coordinator which set a path for the arm manipulator. A new command from the dispatcher sends path information to the motion controller in terms of end points, constraint surface and performance criteria. It also initializes the force sensor and proximity sensor control for grasp activities. The vision coordinator is then switched to a monitoring mode for navigation control, and so on.

The language translation process $L_0 \to L_C \to L_E$, is depicted in Fig. 6.8.

6.6 PERFORMANCE AND ENTROPY

The PNT can be evaluated in **real-time** by testing the computational complexity of their operation which may be expressed uniformly in terms of entropy. Feedback information is communicated to the coordination level from the execution level during the execution of the applied command. Each coordinator, when accessed, issues a number of commands to its associated execution

devices (at the execution level). Upon completion of the issued commands feedback information is received by the coordinator and is stored in the **short-term memory** of the coordination level (Mittman 1990).

This information is stored in the short-term memory of the coordination level. This information is used by other coordinators if necessary, and also to calculate the individual, accrued and overall accrued costs related to the coordination level. Therefore, the feedback information from the execution to the coordination level will be called *on-line, real-time* feedback information.

The performance estimate and the associated subjective probabilities are updated after the k_{ij}-th execution of a task $[(u_t,x_t)_i,S_j]$ and the measurement of the estimate of the observed cost J_{ij}, (Fig. 6.9):

$$J_{ij}(k_{ij}+1) = J_{ij}(k_{ij})+\beta(k_{ij}+1)[J_{obs}(k_{ij}+1)-J_{ij}(k_{ij})] \tag{6.4}$$

$$P_{ij}(k_{ij}+1) = P_{ij}(k_{ij})+\mu(k_{ij}+1)[\Gamma_{ij}(k_{ij}+1)-P_{ij}(k_{ij})]$$

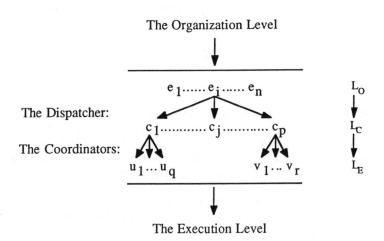

Fig. 6.8 The Language Translation Process

where

$$\Gamma_{ij} = \begin{cases} 1 & \text{if } J_{ij} = \text{Min} \\ 0 & \text{elsewhere} \end{cases}$$

and β and μ are harmonic sequences. Convergence of this algorithm is proven in Saridis and Graham (1984).

The **learning process** is measured by the entropy associated to the subjective probabilities. If

$$H(M) = H(E) + H(T/E) \tag{6.5}$$

where $H(E)$ is the environmental uncertainty and $H(T/E)$ is the pure translation uncertainty. Only the last term can be reduced by learning. The learning curves are given in Fig. 6.10.

6.7 REMARKS

A Petri Net Transducer design of the Coordination level was presented. From the references mentioned in this Chapter it is obvious that other techniques, like Decision schemata have been used in implementing this level.

6.8 REFERENCES

McInroy J.E., Saridis G.N.,(1991), "Reliability Based Control and Sensing Design for Intelligent Machines", in *Reliability Analysis* ed. J.H. Graham, Elsevier North Holland, N.Y.

Mittman M.,(1990), "TOKENPASSER: A Petri Net Specification Tool" *Master's Thesis,* Rensselaer Poly. Inst., Dec.

Nilsson, N.J. (1969), "A Mobile Automaton: An Application of Artificial Intelligence Techniques", *Proc. Int. Joint Conf. on AI*, Washington, D.C.

Peterson, J.L. (1977), "Petri-Nets", *Computing Survey*, 9, No. 3, pp. 223-252, September.

Saridis, G. N. (1983), "Intelligent Robotic Control", *IEEE Trans. on AC*, 28, 4, pp. 547-557, April.

Saridis, G.N. (1985), "Control Performance as an Entropy", *Control Theory and Advanced Technology*, 1, 2, pp. 125-138, Aug.

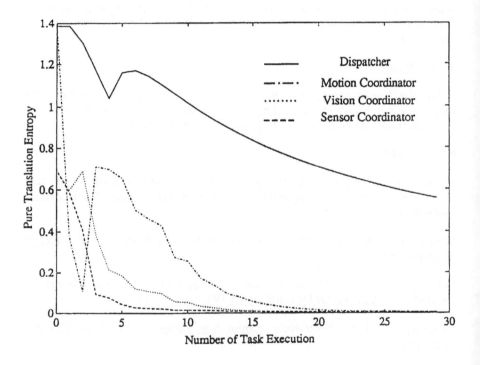

Fig. 6.9 Translation Entropies of the Dispatcher and Coordinators

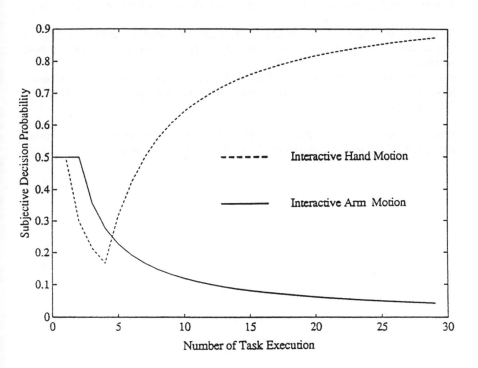

Fig. 6.10 The Learning Curves of Transitions in the Dispatcher

Saridis G. N., (2001) *Entropy in Control Engineering*, World Scientific Publishers, Singapore.

Saridis, G.N. and Graham, J.H. (1984), "Linguistic Decision Schemata for Intelligent Robots", *AUTOMATICA the IFAC Journal*, 20, No. 1, pp. 121-126, Jan.

Saridis, G.N. and Valavanis, K.P. (1988), "Analytical Design of Intelligent Machines", *AUTOMATICA the IFAC Journal*, 24, No. 2, pp. 123-133, March.

M. N. Varvatsoulakis, G.N.Saridis, P. N. Paraskevopoulos,(1998), "A Model for the Organization Level of Intelligent Machines" *Proceedings of the 1998 International Conference of Robotics and Automation,* Leuven Belgium May

M. N. Varvatsoulakis, G. N. Saridis, P. N. Paraskevopoulos, (2000), "Intelligent Organization for Flexible Manufacturing" *IEEE Transactions on Robotics and Automation* Vol.16, No. 2, pp. 180-189, April.
Wang, F. and Saridis, G.N. (1988), "A Model for Coordination of Intelligent Machines Using Petri Nets", *Symposium on Intelligent Control*, Washington, D.C., August.

Wang, F., Saridis, G.N. (1990) "A Coordination Theory for Intelligent Machines" *AUTOMATICA the IFAC Journal*, 35, No. 5, pp. 833-844, Sept.

Chapter 7.

HIERARCHICALLY INTELLIGENT CONTROL: THE EXECUTION LEVEL

7.1 INTRODUCTION

The **Execution level** contains all the hardware required by the Intelligent Machine to execute a task. There is a one-to-one correspondence between hardware groups and coordinators. Therefore their structure is usually fixed. This level also contains all the drivers, VME buses, short memory units, processors, actuators and special purpose devices needed for the execution of a task. After the successful completion of a job feedback information is gene-rated at this level for evaluation and parameter updating of the whole machine. Complexity dominates the performance of this level. Since **precision** is proportional to **complexity**, it also defines the amount of effort required to execute a task. It has been shown that all the activities of this level can be measured by entropy, which may serve as a measure of complexity as well. Minimization of local complexity through feedback, may serve as local design procedure.

The localization of data exchange at this level provides a means of efficient remote control of the Intelligent Machine.

Because of the diversity of the hardware in a general purpose Intelligent Machine, this work will focus on the special case of a robot designed for space construction like the CIRSSE transporter. The following hardware groups are available:

The Vision and Sensory System. This systems consists of two cameras fixed at the ceiling of the lab., two penlight cameras on the wrist of one PUMA arm, and a lazer rangefinder. They are all controlled by a Datacube with a versatile menu of various hardwired functions and a VME bus for internal communications. The functions assigned to them, e.g. create a world model in Cartesian space, find the fiducial marks on the object to be manipulated, or track a moving object are supported by software specialized for the hardware of the system. Calibration and control of the hardware is an important part of the system. Since we are dealing with information processing the system's performance can be easily measured with entropy. Actual data for visual servoing can be generated on the VME bus and transmitted through the Dispatcher to the Motion Control system. Direct connection of the VME bus with the Motion Control System is planned in the future.

The Motion Control System. This system is a unified structure for cooperative motion and force control for multiple arm manipulation. Since motion affects force but not vice versa, motion control is designed independent of the constraint forces, and force control by treating inertial forces as disturbance. Integral force feedback is used with full dynamics control algorithms. The resulting system, named CTOS, was developed as a multiple-processor, VME-bus based, real time robot control system for the CIRSSE 18-degree-of-freedom transporter. It hierarchically integrates the execution algorithms in planning, interaction, and servo control. It works together with the VXWORKS software and provides most of the transformations, and other kinematics and dynamics tools needed for servoing and manipulation. In earlier work it was shown that the control activities can be measured by entropy. Therefore the measure of performance of the Motion Control System is consistent with the rest of the architecture of the Intelligent Machine.

The Grasping System. This system is planned to be separate from the Motion Control System. It would involve the grasping operations, the information gathering from various proximity sensors, and integration of these activities with the gripper motion control. It will be driven by a special coordinator, and provide information back of proper grasping for job control purposes. However at the present time it is only a subsystem of the Motion Control System and follows commands issued by the its Coordinator, for purposes of expediency.

7.2 THE THEORY OF GLOBAL ENTROPY

When we consume energy in order to accomplish some work in our environment we simultaneously generate a low quality residual energy, the cost of which irreversibly reduces the quality of the environment and leads to a chaotic situation. An infinite number of paradigms exist in our environment, starting with the pollution of the air, the water resources, traffic congestion, financial disasters, unemployment with the resulting crime, and in general the decay of the life-sustaining resources of mankind as mentioned by Rifkin (1989).

This low quality energy was discovered by the physicist Clausius which appeared in the second law of thermodynamics and was named Entropy. According to this law the production of work is followed by the production of residual energy that irreversibly increases the total level of the lower level energy and would potentially lead to thermal death.

A different interpretation of entropy was given by Claude Shannon (1963), as a measure of uncertainty in information theory related to communication systems. This interpretation was used by Saridis (1995), to introduce a theory which presents Automatic Control as a generalization of the theory of entropy, based on the designer's uncertainty to obtain the optimal solution. This concept is hereby extended to cover subjects related to the environment, finances, pollution and other problems that puzzle our present society.

7.3 ENTROPY FORMULATION OF MOTION CONTROL

The cost of control at the hardware level can be expressed as an entropy which measures the uncertainty of selecting an appropriate control to execute a task. By selecting an optimal control, one minimizes the entropy, e.g., the uncertainty of execution. The entropy may be viewed in the respect as an energy in the original sense of Boltzmann, as in Saridis (1988).

Optimal control theory utilizes a non-negative functional of the state of the system $x(t) \; \varepsilon \; \Omega_x$ the state space, and a specific control $u(x,t) \; \varepsilon \; \Omega_u \times T; \; \Omega_u \subset \Omega_x$ the set of all admissible feedback controls, to define the performance measure for some initial conditions $x_0(t_0)$, representing a generalized energy function, of the form:

$$V(x_0,t_0) = E\{ \int_{t0}^{tf} L(x,t;u(x,t)) \, dt\} \tag{7.1}$$

where $L(x,t;u(x,t)) > 0$, subject to the differential constraints dictated by the underlying process

$$\begin{aligned} dx/dt &= f(x,u(x,t),w,t); & x(t_0) &= x_0 \\ z &= g(x,v,t); & x(t_f) &\in M_f \end{aligned} \tag{7.2}$$

where x_0, $w(t)$, $v(t)$ are random variables with associated probability densities $p(x_0)$, $p(w(t))$, $p(v(t))$ and Mf a manifold in Ω_x. The trajectories of the system (4.2) are defined for a fixed but arbitrarily selected control $u(x,t)$ from the set of admissible feedback controls Ω_u.

In order to express the control problem in terms of an entropy function, one may assume that the performance measure $V(x_0,t_0,u(x,t))$ is distributed in u according to the probability density $p(u(x,t))$ of the controls $u(x,t) \in \Omega_u$. *The differential entropy* H(u) corresponding to the density is defined as

$$H(u) = - \int_{\Omega u} p(u(x,t)) \ln p(u(x,t)) \, dx$$

and represents the uncertainty of selecting a control $u(x,t)$ from all possible admissible feedback controls Ω_u. The optimal performance should correspond to the maximum value of the associated density $p(u(x,t))$. Equivalently, the optimal control $u^*(x,t)$ should minimize the entropy function H(u).

This is satisfied if the density function is selected to satisfy **Jaynes' Principle of Maximum Entropy** (1956), e.g.,

$$p(u(x,t)) = \exp\{-\lambda - \mu V(x_0,t_0;u(x,t))\} \tag{7.3}$$

where λ and μ are normalizing constants.

It was shown by Saridis (1981), that the expression H(u) representing the entropy for a particular control action u(x,t) is given by:

$$H(u) = \int_{\Omega u} p(x,t;u(x,t))V(x_0,t_0;u(x,t)) \, dx = \lambda + \mu V(x_0,t_0;u(x,t)) \tag{7.4}$$

This implies that the average performance measure of a feedback control problem corresponding to a specifically selected control, is an entropy function. The optimal control $u^*(x,t)$ that minimizes $V(x_0,t_0;u(x,t))$, maximizes $p(x,t;u(x,t))$, and consequently minimizes the entropy H(u).

$$u^*(x,t) : E\{V(x_0,t_0;u^*(x,t))\} = \min\int_{\Omega u} V(x_0,t_0;u(x,t))p(u(x,t))dx \tag{7.5}$$

This statement is the generalization of a theorem proven by Saridis (1981), and establishes equivalent measures between information theoretic and optimal control problem and provides the information and feedback control theories with a common measure of performance.

7.4 ENTROPY MEASURES OF STEREO VISION SYSTEMS

The optimal control theory designed mainly for motion control, can be implemented for vision control, path planning and other sensory system pertinent to an Intelligent Machine by slightly modifying the system equations and cost functions. After all one is dealing with real-time dynamic systems which may be modeled by a dynamic set of equations.

A **Stereo Vision** system of a pair of cameras mounted at the end of a robot arm, may be positioned at i=1,..N different view points to reduce problems with noise, considered one at a time due to time limitations. The accuracy of measuring the object's position depends upon its relative position in the camera frame. Consequently, each viewpoint will have different measurement error and time statistics. These statistics may be generated to define the uncertainty of the measurement of the Vision system as in McInroy and Saridis (1991).

For a point c of the object, the measurement error of its 3-D position in the camera coordinate frame e_{pc} is given by:

$$e_{pc} = M_c \, n_c \tag{7.6}$$

where n_c is the 3-D image position errors, and M_c an appropriate 3X3 matrix, depending on the position of the object.

The linearized orientation error is given by:

$$\delta = (M^TM)^{-1}M^TM'Fn \tag{7.7}$$

where

δ is the orientation error in the camera frame, M is a matrix formed from camera coordinate frame positions,
M' is a constant matrix,
F is the matrix formed from the camera parameters and measured positions,
n is the vector of the image position errors at the four points.

A vector containing the position and orientation errors due to image noise is given by:

$$e_c = [e^T_{pc}\delta^T]^T = Ln \tag{7.8}$$

where L depends on the camera parameters and the four measured camera frame positions of the points. The statistics of the image noise n, due to individual pixel errors are assumed to be uniformly distributed. Assuming that feature matching centroids is used by the vision system, its distributions tend to be independent Gaussian, due to the Central Limit Theorem.

$$n \approx N(0,C_v) \text{ and } e_c \approx N(0,LC_vL^T) \tag{7.9}$$

The time which each vision algorithm consumes is also random due to the matching period. Therefore the total vision time, for the ith algorithm that includes camera positioning time, image processing time, and transformation to the base frame, is assumed Gaussian:

$$t_{vi} \approx N(\mu_{tvi},\sigma^2_{tvi}). \tag{7.10}$$

Once the probability density functions are obtained, the resulting Entropies $H(t_{vi})$, and $H(e_c)$, are obtained in a straight forward manner for the ith Algorithm (McInroy and Saridis 1991):

$$H(t_{vi}) = \ln\sqrt{2\pi e\sigma^2_{tvi}} \tag{7.11}$$
$$H(e_c) = \ln\sqrt{(2\pi e)^6 \det[C_v]} + E\{\ln[\det L_i]\}$$

The total Entropy, may be used as a measure of uncertainty of the Vision system (imprecision), and can be minimized with respect to the available system parameters:

$$H(V) = H(t_{vi}) + H(e_c).$$ (7.12)

7.5 REMARKS

The formulation of the Execution level was stated using Entropy measures. Entropy was also used to measure sensing and vision activities of the system.

7.6 REFERENCES

Boltzmann L. (1872), "Further Studies on Thermal Equilibrium Between Gas Molecules", *Wien Ber.*, Vol. *66*, p. 275.

Feld'baum, A.A. (1965), *Optimal Control Systems*, Academic Press, New York.

Jaynes, E.T. (1957), "Information Theory and Statistical Mechanics", *Physical Review*, Vol.4, pp. 106.

Kumar, P.R., Varaiya, P., (1986), *Stochastic Systems; Estimation, Identification and Adaptive Control*, Prentice Hall, Englewood Cliffs, NJ.

Prigogine I.,(1980), *From Being to Becoming* W. H. Freeman & Co. San Francisco, CA.

Kumar, P.R., Varaiya, P., (1986), *Stochastic Systems: Estimation, Identification, and Adaptive Control*, Prentice Hall Inc. Englewood Cliffs, NJ.

McInroy J.E., Saridis G.N.,(1991), "Reliability Based Control and Sensing Design for Intelligent Machines", in *Reliability Analysis* ed. J.H. Graham, Elsevier North Holland, N.Y.

Morse, A.S., (1990), "Towards a Unified Theory of Parameter Adaptive Control: Tunability", *IEEE Trans. on Automatic Control*, Vol. 35, No. 9, pp. 1002-1012, September.

Morse, A.S., (1992), "Towards a Unified Theory of Parameter Adaptive Control-Part II:Certainty Equivalence and Implicit Tuning", *IEEE Trans. on Automatic Control*, Vol. 37, No. 1, pp. 15-29, January.

Saridis, G.N. (1977), *Self-Organizing Controls of Stochastic Systems*, Marcel Dekker, New York, New York.

Saridis, G.N. (1979), "Toward the Realization of Intelligent Controls", *IEEE Proceedings*, Vol. 67, No. 8.

Saridis, G. N. (1983), "Intelligent Robotic Control", *IEEE Trans. on Automatic Control*, Vol. 28, No. 4, pp. 547-557, April.

Saridis, G. N., (1985a), "Intelligent Control: Operating systems in Uncertain Environments", Chapter 7 in *Uncertainty and Control*, Ed. J. Ackermann, Springer Verlag, Berlin pp. 215-233.

Saridis, G.N. (1985b), "An Integrated Theory of Intelligent Machines by Expressing the Control Performance as an Entropy", *Control Theory and Advanced Technology*, Vol. 1, No. 2, pp. 125-138, Aug.

Saridis, G.N. (1985c), "Foundations of Intelligent Controls", *Proceedings of IEEE Workshop on Intelligent Controls*, p. 23, RPI, Troy, New York, NY.

Saridis, G.N. (1988), "Entropy Formulation for Optimal and Adaptive Control", *IEEE Transactions on Automatic Control*, Vol. 33, No. 8, pp. 713-721, Aug.

Saridis, G.N. (1989), "Analytic Formulation of the IPDI for Intelligent Machines", *AUTOMATICA the IFAC Journal*, 25, No. 3, pp. 461-467.

Saridis, G.N. and Graham, J.H. (1984), "Linguistic Decision Schemata for Intelligent Robots", *AUTOMATICA the IFAC Journal*, 20, No. 1, pp. 121-126, Jan.

Saridis, G.N. and Lee, C.S.G. (1979), "Approximation of Optimal Control for Trainable Manipulators", *IEEE Trans. on Systems Man and Cybernetics*, Vol.8, No. 3, pp. 152-159, March.

Saridis, G.N. and Valavanis, K.P. (1988), "Analytical Design of Intelligent Machines", *AUTOMATICA the IFAC Journal*, 24, No. 2, pp. 123-133, March.

Shannon, C. and Weaver, W. (1963), *The Mathematical Theory of Communications*, Illini Books.

Chapter 8.

HIERARCHICALLY INTELLIGENT CONTROL: APPLICATION TO ROBOTIC SYSTEMS

8.1 INTRODUCTION

The theory of Intelligent Controls has direct application to the design of Intelligent Robots. The IPDI provides a means of structuring hierarchically the levels of the machine. Since for a passive task the flow of knowledge through the machine must be constant, it assigns the highest level with the highest machine intelligence and smallest complexity (size of data base), and the lowest level with the lowest machine intelligence and largest complexity. Such a structure agrees with the concept of most organizational structures encountered in human societies. Application to machine structures is straight forward.

Even at the present time there is a large variety of applications for intelligent machines. Automated material handling and assembly in an automated factory, automation inspection, sentries in a nuclear containment are some of the areas where intelligent machines have and will find a great use. However, the most important application for the author's group is the application of Intelligent Machines to unmanned space exploration where, because of the distance involved, autonomous anthropomorphic tasks must be executed and only general commands and reports of executions may be communicated. Such tasks are suitable for intelligent robots capable of executing anthropomorphic tasks in unstructured uncertain environments.

They are structured usually in a human-like shape and are equipped with vision and other tactile sensors to sense the environment, two areas to execute tasks and locomotion for appropriate mobility in the unstructured environment. The controls of such a machine are performed according to the Theory of Hierarchically Intelligent Machines discussed in previous Chapters.

The three levels of controls, obeying the **Principle of Increasing Precision with Decreasing Intelligence**, have been tested on a test bed composed of two PUMA 600 robot arms with stereo vision and force sensing, built in the Center for Intelligent Robotics for Space Exploration at Rensselaer Polytechnic Institute(Fig. 8.1).

8.2 THE ARCHITECTURE OF THE ORGANIZATION LEVEL
In this case, and for practical reasons, the organization level of this structure has been replaced by a human operator; and there is no analytic model for this level.

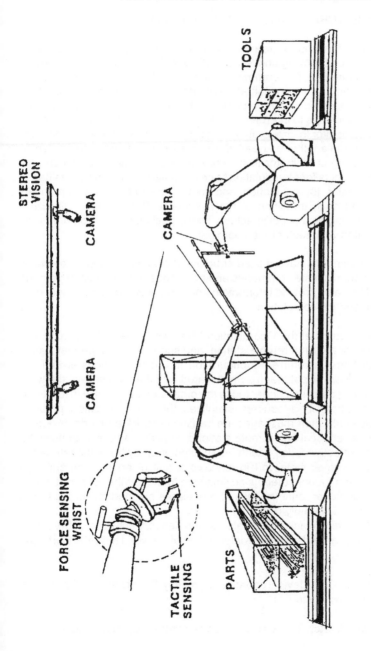

Fig. 8.1 Hierarchically Intelligent Control of Manipulators with Sensory Feedback as Developed at the CIRRSE Laboratory of the Rensselaer Poly. Institute

8.3 THE ARCHITECTURE OF THE COORDINATION LEVEL

The *Coordination level* is a tree structure of *Petri Net Transducers* as coordinators, proposed by Wang and Saridis (1990) with the Dispatcher as the root and is shown in Figure 8.1. The commands generated by the Human Organizer defining every specific plan is transmitted, asynchronously, to the Dispatcher of the Coordination level, modeled by a reconfigurable Petri Net Transducer (PNT). The function of the Dispatcher is to interpret the plan and assign individual tasks to the other coordinators, monitor their operation, and transmit messages and commands from one coordinator to another as needed. As an example, a command is sent to the vision and sensing coordinator to generate a model of the environment, the coordinates of the objects for manipulation to be tabulated, and then transmitted to the motion coordinator for navigation and motion control. This command is executed by having each transition of the associated Petri Nets to initialize a package corresponding to a specific action. These packages are stored in short memories associated with each of the coordinators.

The rest of the coordinators have a fixed structure with alternate menus available at request. They communicate commands and messages with each other, through the Dispatcher. They also provide information about reception of a message, data memory location, and job completion.

No data is communicated at the Coordination level, since the task planning and monitoring may be located in a remote station, and such an exchange may cause a channel congestion. A preferred configuration for such situations is that the coordinators with a local dispatcher may be located with the hardware at the work site, while a remote dispatcher, connected to the organizer, interacts with local one from a remote position. This concept simplifies the communication problem considerably, since only short messages are transmitted back and forth through a major channel between local and remote stations, requiring a narrow bandwidth. An example of the effectiveness of such an architecture may be demonstrated in space construction, where robots work in space while task planning and monitoring is done on earth. Even though, there is no limitation to the number of coordinators attached to the Dispatcher, only the following ones are planned for an Intelligent Robot for space truss assembly (Fig. 8.2).

Vision Coordinator. This device coordinates all the sensory activities of the robot, with cameras and lazers, and generates information of the world model in Cartesian coordinates.

Sensory Coordinator. This device coordinates mainly the activities of proximity sensors, for fine motion control.

Motion Control Coordinator. This device receives control, object and obstacle

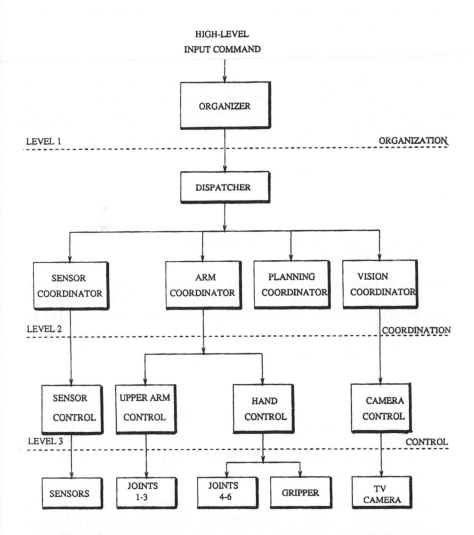

Fig. . 8.2 Hierarchically Intelligent Control for a Manipulator with Sensory Feedback

information and uses it to navigate and move multiple robotic arms and other devices, for object manipulation and task execution. It also assigns the appropriate operations on the data acquired for the desired application.

Entropy measures, expressing the system reliability, are developed by McInroy and Saridis (1991), at each coordinator such that they may be used to minimize the complexity and improve the reliability of the system.

8.4 THE ANALYTIC MODEL

Petri nets have been proposed in Chapter 3, as possible devices to communicate and control complex heterogenous processes. These nets provide a communication protocol among stations of the process as well as the control sequence for each one of them. Abstract task plans, suitable for many environments are generated at the organization level by a grammar created by Wang and Saridis (1990):

$$G = (N, \Sigma_o, P, S)$$

where

$N = \{S, M, Q, H\}$ = Non-terminal symbols
$\Sigma_o = \{A_1, A_2,...A_n\}$ = Terminal Symbols (activities)
P = Production rules

Petri Net Transducers (PNT) proposed first by Wang and Saridis, are Petri net realizations of the **Linguistic Decision Schemata** introduced by Saridis and Graham (1984), as linguistic decision making and sequencing devices. They are defined as 6-tuples:

$$M = (N, \Sigma, \delta, G, \mu, F)$$

where

$N = (P, T, I, O)$ = A Petri net with initial marking μ,
Σ = a finite input alphabet
δ = a finite output alphabet
σ = a translation mapping from $T \times (\Sigma \cup \{\lambda\})$ to finite
  sets of δ^* and $F \subset R(\mu)$ a set of final markings.

A **Petri Net Transducer** (PNT) is a Petri Net with input and output tapes. It communicates with the other levels through input and output languages called **Petri Net Languages (PNL)**. In addition to its on-line decision making capability PNT's have the potential of generating communication protocols, learning by feedback, ideal for the communication and control of coordinators and their dispatcher in real time. Their architecture for the particular Truss Construction in space is given in Fig. 8.3, and may follow a scenario suitable for the implementation of an autonomous intelligent robot. Decision making in the coordination structure is accomplished by **Task Scheduling** and **Task Translation** (Wang and Saridis 1990).

The sequence of events transmitted from the organization level is received by the dispatcher which requests a world model with coordinates from a vision coordinator.

The vision coordinator generates appropriate database and upon the dispatcher's command communicates it to the planning coordinator which set a path for the arm manipulator. A new command from the dispatcher sends path information to the motion controller in terms of end points, constraint surface and performance criteria. The vision coordinator is then switched to a monitoring mode for navigation control, and so on.

The PNT can be evaluated in real-time by testing the computational complexity and system reliability of their operation which may be expressed uniformly in terms of entropy.

Feedback information is communicated to the coordination level from the execution level during the execution of the applied command. Each coordinator, when accessed, issues a number of commands to its associated execution devices (at the execution level). Upon completion of the issued commands feedback information
is received by the coordinator and is stored in the **short-term memory** of the

This information is used by other coordinators if necessary, and also to calculate the individual, accrued and overall accrued costs related to the coordination level. Therefore, the feedback information from the execution to the coordination level will be called **on-line, real-time** feedback information, and has learning properties (Lima and Saridis 1996).

The performance estimate and the associated subjective probabilities are updated after the k_{ij}-th execution of a task $[(u_t,x_t)_i,S_j]$ and the measurement of the estimate of the observed cost J_{ij}:

$$J_{ij}(k_{ij}+1) = J_{ij}(k_{ij})+\beta(k_{ij}+1)[J_{obs}(k_{ij}+1)-J_{ij}(k_{ij})] \qquad (8.1)$$

$$P_{ij}(k_{ij}+1) = P_{ij}(k_{ij})+\mu(k_{ij}+1)[\Gamma_{ij}(k_{ij}+1)-P_{ij}(k_{ij})]$$

where

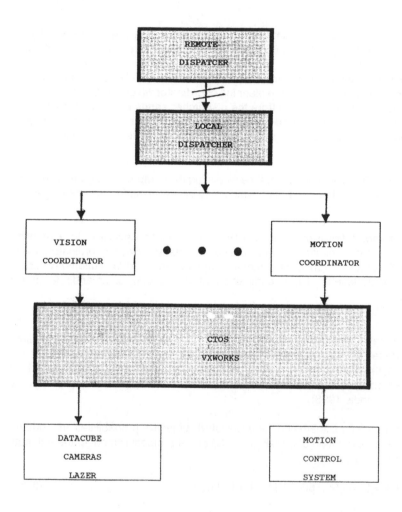

Fig. 8.3 Telerobotics Testbed Configuration Coordination Level

$$\Gamma_{ij} = \begin{cases} 1 & \text{if } J_{ij} = \text{Min} \\ 0 & \text{elsewhere} \end{cases}$$

and β and μ are harmonic sequences. Convergence of this algorithm is proven in Saridis and Graham (1984).

The **learning process** is measured by the entropy associated to the subjective probabilities. If

$$H(C) = H(E) + H(T/E) \tag{8.2}$$

where $H(E)$ is the environmental uncertainty and $H(T/E)$ is the pure translation uncertainty. Only the last term can be reduced by learning.

8.5 THE ARCHITECTURE OF THE EXECUTION LEVEL

The **Execution level** contains all the hardware required by the Intelligent Machine to execute a task. There is a one-to-one correspondence between hardware groups and coordinators. Therefore their structure is usually fixed. This level also contains all the drivers, VME buses, short memory units, processors, actuators and special purpose devices needed for the execution of a task. After the successful completion of a job feedback information is generated at this level for evaluation and parameter updating of the whole machine. Complexity dominates the performance of this level. Since **precision** is proportional to **complexity**, it also defines the amount of effort required to execute a task. It has been shown that all the activities of this level can be measured by entropy, which may serve as a measure of complexity as well. Minimization of local complexity through feedback, may serve as local design procedure.

The localization of data exchange at this level provides a means of efficient remote control of the Intelligent Machine.

The diversity of the hardware in a general purpose Intelligent Machine is too abstract for the Truss Construction project under consideration. Therefore, this work will focus on the special case of a robot designed for space construction like the CIRSSE transporter. The following hardware groups shown in the Laboratory configuration of Fig. 8.4, are available:

The Vision System. This systems consists of two cameras fixed at the ceiling of the lab., two penlight cameras on the wrist of one PUMA arm, and a lazer range finder. They are all controlled by a Datacube with a versatile menu of various hardwired

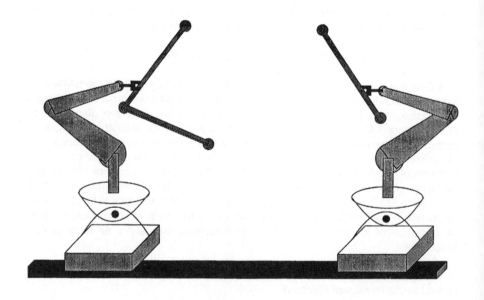

Fig. 8.4 A Case Study of Robotic Truss Assembly

functions and a VME bus for internal communications. The functions assigned to them, e.g. create a world model in Cartesian space, find the fiducial marks on the object to be manipulated, or track a moving object are supported by software specialized for the hardware of the system. Calibration and control of the hardware is an important part of the system. Since we are dealing with information processing the system's performance can be easily measured with entropy. Actual data for visual servoing can be generated on the VME bus and transmitted through the Dispatcher to the Motion Control system. Internal communication on a VME bus, and direct connection with the VME bus of the Motion Control System has been developed as the VSS system.

The Sensory System. This system consists of two proximity sensors that are used for control of fine motion of the manipulator and the gripper. It functions in a manner very similar to the vision system, but it acts independent of the Motion Control and Vision System busses.

The Motion Control System. This system is a unified structure for cooperative motion and force control for multiple arm manipulation. Since motion affects force but not vice versa, motion control is designed independent of the constraint forces, and force control by treating inertial forces as disturbance. Integral force feedback is used with full dynamics control algorithms. The resulting system, named CTOS, was developed as a multiple-processor, VME-bus based, real time robot control system for the CIRSSE 18-degree-of-freedom transporter. It hierarchically algorithms in planning, interaction, and servo control. It works together with the VXWORKS software and provides most of the transformations, and other kinematics and dynamics tools needed for servoing and manipulation. In earlier work it was shown that the control activities can be measured by entropy (Saridis 1988). Therefore the measure of performance of the Motion Control System is consistent with the rest of the architecture of the Intelligent Machine.

8.6 ENTROPY FORMULATION OF MOTION CONTROL

The cost of control at the hardware level can be expressed as an entropy which measures also the uncertainty of selecting an appropriate control to execute a task. By selecting an optimal control, one minimizes the entropy, e.g., the uncertainty of execution. The entropy may be viewed in this respect as an irreversible lower level energy in the original sense of Boltzmann (1872). The appropriate task for the Truss Assembly system is depicted in Fig. 8.4.

Optimal control theory utilizes the non-negative cost functional of the state of the system $x(t) \varepsilon \Omega_x$ the state space, and a specific control $u(x,t) \varepsilon \Omega_u \times T$; $\Omega_u \subset \Omega_x$ the set of all admissible feedback controls, to define the performance measure for some initial conditions $x_0(t_0)$, representing a generalized energy function, of the form:

$$V(x_0, t_0) = E\{ \int_{t_0}^{\overline{t}} L(x, t; u(x, t)) \, dt\}$$ (8.3)

where $L(x, t; u(x, t)) > 0$, subject to the differential constraints, dictated by the underlying process

$$dx/dt = f(x, u(x, t), w, t); \quad x(t_0) = x_0$$
$$z = g(x, v, t); \qquad x(t_r) \in M_f$$ (8.4)

where x_0, $w(t)$, $v(t)$ are random variables with associated probability densities $p(x_0)$, $p(w(t))$, $p(v(t))$ and M_f a manifold in Ω_x. The trajectories of the system eq. (8.4) are defined for a fixed but arbitrarily selected control $u(x, t)$ from the set of admissible feedback controls Ω_u.

In order to express the control problem in terms of an entropy function, one may assume that the performance measure $V(x_0, t_0, u(x, t))$ is distributed in u according to the probability density $p(u(x, t))$ of the admissible controls $u(x, t) \in \Omega_u$. **The differential entropy** $H(u)$ corresponding to the density is defined as

$$H(u) = -\int_{\Omega_u} p(u(x, t)) \ln p(u(x, t)) \, dx$$

and represents the uncertainty of selecting a control $u(x, t)$ from all possible admissible feedback controls Ω_u. The optimal performance should correspond to the maximum value of the associated density $p(u(x, t))$. Equivalently, the optimal control $u^*(x, t)$ should minimize the entropy function $H(u)$. This is satisfied if the density function is selected to satisfy **Jaynes' Principle of Maximum Entropy** (Jaynes 1957), e.g.,

$$p(u(x, t)) = \exp\{-\lambda - \mu V(x_0, t_0; u(x, t))\}$$ (8.5)

where λ and μ are normalizing constants.

It was shown by Saridis (1988), that the expression $H(u)$ representing the entropy for a particular control action $u(x, t)$ is given by:

$$H(u) = \int_{\Omega_u} p(x, t; u(x, t)) V(x_0, t_0; u(x, t)) \, dx = \lambda + \mu V(x_0, t_0; u(x, t))$$ (8.6)

This implies that the average performance measure of a feedback control problem corresponding to a specifically selected control, is an entropy function. The optimal control $u^*(x, t)$ that minimizes $V(x_0, t_0; u(x, t))$, maximizes $p(x, t; u(x, t))$, and consequently minimizes the entropy $H(u)$.

$$u^*(x,t) : E\{V(x_0,t_0;u^*(x,t))\} = \min\int_{\Omega_u} V(x_0,t_0;u(x,t))p(u(x,t))dx \qquad (8.7)$$

This statement is the generalization of a theorem proven by Saridis (1988), and establishes equivalent measures between information theoretic and optimal control problem and provides the information and feedback control theories with a common measure of performance.

8.7 ENTROPY MEASURE OF THE VISION SYSTEM

The Intelligent Control theory, designed mainly for motion control, can be implemented for vision control, path planning and other sensory system pertinent to an Intelligent Machine by slightly modifying the system equations and entropy functions as in Chapter 3. After all one is dealing with real-time dynamic systems which may be modeled by a dynamic set of equations.

A **Stereo Vision** system of a pair of cameras mounted at the end of a robot arm, may be positioned at $l=1,..N$ different view points to reduce problems with noise, considered one at a time due to time limitations. The accuracy of measuring the object's position depends upon its relative position in the camera frame. Consequently, each viewpoint will have different measurement error and time statistics. These statistics may be generated to define the uncertainty of the measurement of the Vision system as in McInroy and Saridis (1991).

For a point c of the object, the measurement error of its 3-D position in the camera coordinate frame e_{pc} is given by:

$$e_{pc} = M_c \, n_c \qquad (8.8)$$

where n_c is the 3-D image position errors, and M_c an appropriate 3X3 matrix, depending on the position of the object.

The linearized orientation error is given by:

$$\delta = (M^TM)^{-1}M^TM'Fn \qquad (8.9)$$

where

δ is the orientation error in the camera frame,
M is a matrix formed from camera coordinate frame positions,
M' is a constant matrix,
F is the matrix formed from the camera parameters and measured positions,
n is the vector of the image position errors at the four points.

A vector containing the position and orientation errors due to image noise is given by:

$$e_c = [e^T_{pc}\delta^T]^T = Ln \tag{8.10}$$

where L depends on the camera parameters and the four measured camera frame positions of the points. The statistics of the image noise n, due to individual pixel errors are assumed to be uniformly distributed. Assuming that feature matching centroid is used by the vision system, its distributions tend to be independent Gaussian, due to the Central Limit Theorem.

$$n \approx N(0,C_v) \quad \text{and} \quad e_c \approx N(0,LC_vL^T) \tag{8.11}$$

The time which each vision algorithm consumes is also random due to the matching period. Therefore the total vision time, for the ith Algorithm that includes camera positioning time, image processing time, and transformation to the base frame, is assumed Gaussian:

$$t_{vi} \approx N(\mu_{tvi},\sigma^2_{tvi}). \tag{8.12}$$

Once the probability density functions are obtained, the resulting Entropies $H(t_{vi})$, and $H(e_c)$, are immediately obtained for the ith Algorithm (McInroy and Saridis 1991):

$$H(t_{vi}) = \ln\sqrt{2\pi e\sigma^2_{tvi}} \tag{8.13}$$
$$H(e_c) = \ln\sqrt{(2\pi e)^6}\det[C_v] + E\{\ln[\det L_i]\}$$

The total Entropy, may be used as a measure of uncertainty of the Vision system (imprecision), and can be minimized with respect to the available system parameters:

$$H(V) = H(t_{vi}) + H(e_c). \tag{8.14}$$

8.8 ENTROPY MEASURE FOR THE SENSORY SYSTEM

The proximity sensory system has similar properties as the vision system. Therefore its entropy measure should be similar to eq. (8.14):

$$H(S) = H(t_{vi}) + H(e_c). \tag{8.15}$$

8.9 TOTAL ENTROPY OF THE SYSTEM

Since the Coordination level works on a different time scale than the Execution level, the total Entropy of the system is obviously the sum of the partial Entropies of the Coordination H(C), the Motion Control H(MC), the Vision system H(V), and the Sensory system H(S).

$$H(TOTAL) = H(C) + H(MC) + H(V) + H(S) \qquad (8.16)$$

By selecting individual controls to minimize the total Entropy H(TOTAL) one obtains the optimal truss assembly procedure for space construction Fig.8.5, is a photograph of the experimental system for Truss Assembly in space designed and operated at the Center for Intelligent Robotic Systems for Space Exploration(CIRSSE) at the Rensselaer Polytechnic Institute in 1993.

8.10 REMARKS

The Truss Assembly in space project, developed at CIRSSE of RPI, was used as a case study, of the application of Intelligent Control. It demonstrated the usefulness of using Entropy as a unifying measure, for optimal control of a process, combining various disjoint disciplines. This presents a complete justification of the theoretical work presented in this book, and other publications

More importantly, it has established that the cost of performance, expressed as entropy, is an irreversible phenomenon in time, since such a construction assumes energy consumption in space flights, visual recognition, decision making and control activities that cannot be reversed in time. It also stands as a link with other contemporary attempts to introduce entropy as the unifying measure in disciplines like biotechnology, ecology, economics and manufacturing.

Finally, as a caveat, it shows the way to the future of the human race to face the problem of overpopulation and over pollution of the earth by colonizing other planets in space. It demonstrates the capabilities of Intelligent Machines to assist humans in this adventure.

8.11 REFERENCES

Boltzmann, L. (1872) "Further Studies on Thermal Equilibrium between Gas Molecules" *Wien Ber.*, **66**, p. 275.

Jaynes, E.T. (1957), "Information Theory and Statistical Mechanics", **Physical Review**, Vol.4, pp. 106.

Lima, P. U., Saridis G.N.,(1996), ***Design of Intelligent Control Systems based on Hierarchical Stochastic Automata.*** World Scientific, Singapore

Fig. 8.5 The CIRSSE Laboratory for Intelligent Robotic Machines

McInroy J.E., Saridis G.N.,(1991), "Reliability Based Control and Sensing Design Intelligent Machines", in *Reliability Analysis* ed. J.H. Graham, Elsevier North Holland, N.Y.

Saridis, G. N. (1983), "Intelligent Robotic Control", *IEEE Trans. on Automatic Control*, Vol. 28, No. 4, pp. 547-557, April.

Saridis, G. N. (1988), "Entropy Formulation for Optimal and Adaptive Control", *IEEE Transactions on Automatic Control*, Vol. 33, No. 8, pp. 713-721, Aug.

Saridis G. N. (1996), "Architectures for Intelligent Controls" in *Intelligent Control Systems*, Edited by Madan. M. Gutta and Nares K. Sinha, Chapter 6, pp.127-148, IEEE Press.

Saridis, G.N. and Valavanis, K.P. (1988), "Analytical Design of Intelligent Machines", *AUTOMATICA the IFAC Journal*, 24, No. 2, pp. 123-133, March.

Wang, F., Saridis, G.N. (1990) "A Coordination Theory for Intelligent Machines" *AUTOMATICA the IFAC Journal*, 35, No. 5, pp. 833-844,Sept.

Valavanis, K.P., Saridis, G.N., (1992), *Intelligent Robotic System Theory: Design and Applications*, Kluwer Academic Publishers, Boston, MA.

Moed M. C., and Saridis G.N., (1990), "A Boltzmann Machine for the Organization of Intelligent Machines" *IEEE Transactions on Systems Man and Cybernetics* Vol. 20 No. 5, Oct. pp.1094-1102

Saridis G. N., and Moed M. C., (1998), "Analytic Formulation of Intelligent Machines as Neural Nets", *Proc. of IEEE Conference on Intelligent Control*, Washington D. C., Aug.

Saridis, G.N., (1996)," Architectures for Intelligent Controls" *Chapter 6*, in *Intelligent Control Systems*, M. M. Gupta, N. K. Singh (eds) IEEE Press New York NY.

Chapter 9.

INTELLIGENT MANUFACTURING

9.1.INTRODUCTION

The evolution of the digital computer in the last thirty years has made possible to develop fully automated systems that successfully perform human dominated functions in industrial, space, energy, biotechnology, office, and home environments, generating waste interpreted as entropy. Therefore, automation has been a major factor in modern technological developments. It is aimed at replacing human labor in

> a. hazardous environments,
> b. tedious jobs,
> c. inaccessible remote locations and
> d. unfriendly environments.

It possesses the following merits in our technological society: reliability, reproducibility, precision, independence of human fatigue and labor laws, and reduced cost of high production.

Modern Intelligent Robotic Systems, using entropy as a measure of performance, are typical applications of Automation to an industrial society (Valavanis, Saridis 1992). They are equipped with means to sense the environment and execute tasks with minimal human supervision, leaving humans to perform higher level jobs. They are also adaptable to the taste of every designer.

Manufacturing on the other hand, is an integral part of the industrial process, and is defined as follows:

> **Manufacturing is to make or process a finished product through a large scale industrial operation.**

In order to improve profitability, modern manufacturing, which is still a **disciplined art**, always involves some kind of automation. Going all the way and fully automating manufacturing is the dream of every industrial engineer. However, it has found several roadblocks in its realization, measured by entropy: environmental pollution, acceptance by the management, loss of manual jobs, marketing vs. engineering. The National Research Council reacted to these problems by proposing a solution which involved among other items a new discipline called: **Intelligent Manufacturing** (The Comprehensive Edge 1989).

Intelligent Manufacturing is the process that utilizes Intelligent Control, with entropy as a measure, in order to accomplish its goal. It possesses several degrees of autonomy, by demonstrating (machine) intelligence to make crucial decisions during the process. Such decisions involve scheduling, prioritization, machine selection, product flow optimization, etc., in order to expedite production and improve profitability.

9.2. INTELLIGENT MANUFACTURING

Intelligent Manufacturing is an immediate application of Hierarchically Intelligent Control discussed in Chapter 4. It has been defined as the combination of disciplines of Artificial Intelligence, Operations Research and Control System Theory (see Fig. 1.1), in order to perform tasks with minimal interaction with a human operator. One of its hierarchical applications, proposed by Saridis (1996), is an architecture based on the **Principle of Increasing Precision with Decreasing Intelligence (IPDI),** which is the manifestation on a machine of the human organizational pyramid The Principle is realized by three structural levels using entropy as a common measure (see Fig. 9.1):

1. The Organization level (Saridis, Moed 1988)
2. The Coordination level (Saridis, Graham 1984, Wang, Saridis 1990)
3. The Execution level (Saridis 1979).

Intelligent Manufacturing can be implemented in the Factory of the Future by modularizing the various workstations and assigning Hierarchically Intelligent Control to each one of them, the following tasks;

1. Product Planning to the Organization level
2. Product Design and Hardware Assignment and Scheduling to the Coordination level
3. Product Generation to the Execution level

The algorithms at the different levels may be modified according to the taste of the designer, and the type of the process. However, manufacturing can be thus streamlined and optimized by minimizing the total entropy of the process. Robotics may be thought as an integral part of Intelligent Manufacturing and be included as part of the workstations. This creates an intelligently versatile automated industrial environment where, every time, each unit may be assigned different tasks by just changing the specific algorithms at each level of the hierarchy (see Fig. 9.2) in contrast to the serial production which requires more equipment and effort (Fig. 9.3).

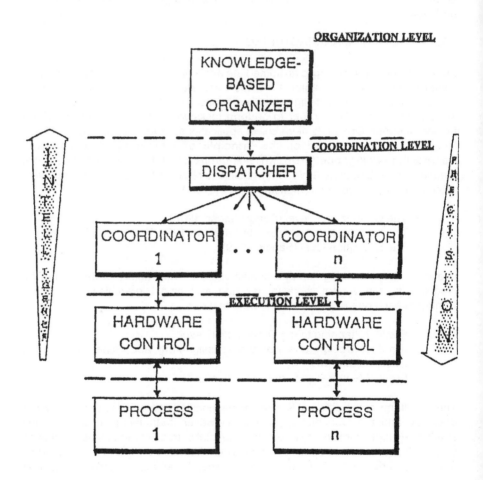

Fig. 9.1 A Complete Hierarchically Intelligent Control System

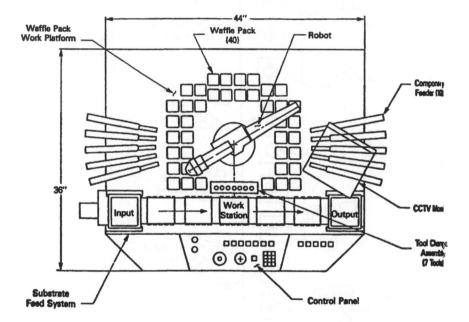

Fig. 9.2 A Circular Design Robotic Assembly Station

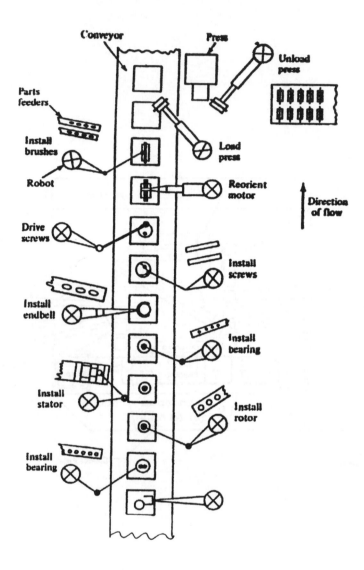

Fig. 9.3 A Linear Design Robotic Assembly Station

This approach is designed to reduce interruptions due to equipment failures, bottlenecks, rearrangement of orders, material delays and other typical problems that deal with production , assembly and product inspection. A case study dealing with a nuclear plant may be found in (Valavanis, Saridis 1992).

At the present time the application of such technology, even though cost-effective in competitive manufacturing, is faced with significant barriers due to (The Comprehensive Edge 1989);

    a. Inflexible organizations
    b. Inadequate available technology
    c. Lack of appreciation and
    d. Inappropriate performance measures

However, international competition, and need for more reliable, precisely reproducible products is directing modern manufacturing towards more sophistication and the concept of an Intelligent Factory of the future. An interesting application based on the work of Varvatsoulakis, Saridis, and Paraskevopoulos (1998) is discussed in the sequel.

9.3 ARCHITECTURE OF INTELLIGENT SCHEDULING FOR PRODUCTION

The theory of Hierarchical Intelligent Control is general and versatile enough to be used to control as well as schedule tasks in Manufacturing. Such is the case of automated multiple product scheduling in a modern factory.

Automated multiple product scheduling is needed when the factory produces more than one product on the same set of stations and the ordering of production must be set as a minimum operating cost scheduling problem. The problem is mathematically formulated to set the order of production using entropy as a measure in the Intelligent Control's three level structure (Varvatsoulakis, Saridis, and Paraskevopoulos 1998, 1999). The complete system is able to issue high-level task plans and use them to control the stations of the factory in the lower level of the hierarchy. The system includes a learning algorithm designed to obtain asymptotically optimal task plans for production control in uncertain environments. A case study in (Varvatsoulakis, Saridis, and Paraskevopoulos 1999) demonstrates the scheduling of the assembly of a gear box. The intelligent manufacturing approach is presented in the following sections.

9.3.1 Product Scheduling Architecture: The Organization Level

The Organization level, discussed in Chapter 4, is intended to perform operations as **Task Representation, Planning, Decision Making, and Learning,** in an abstract way and produce strings of commands to be executed by the lower levels of the system. A Boltzmann type of neural net was found to suffice to produce these operations. The analytic model proposed for multiple product scheduling in manufacturing follows the pattern of hierarchically intelligent control and is divided in three levels.

The organization level, (Fig.9.4), that sets up the task planning is composed of the following subtasks:

Task Representation is the association of a station node to a number of tasks. The probability of its activation, and the weight for transfer are assigned and its associated entropy is calculated. Define:

- The ordered set of sublevels $L=\{l_1,...l_k\}$ is the set of abstract primitive subtasks, each one containing number nl_i, $l=1,...k$ of independent primitive nodes.

- The set of nodes $D=\{d_{i1},...d_{i,nl}, l=1,...k\}$, is the subtask domain with each node containing number of primitive objects; for convenience they are represented by their subscripts.

- The set $B \subseteq D$ contains the starting nodes.

- The set $S \subseteq D$ contains the terminal nodes

- The set of random variables $Q=\{q_{i1},...q_{i,ni}; l=1,...k\}$ with binary values $[0,1]$, represents the inactive or active state of events associated with the nodes D.

- The set of probabilities $P=\{p_{i1},...p_{i,ni}; l=1,...k\}$ associated with the random variables Q, as follows:

$$P=\{p_{ij}=Prob(q_{ij}=1); l=1,...k; j=1,...nl$$

Task planning is the ordering of the production activities, obtained by properly concatenating the appropriate abstract primitive nodes $\{d_{m,li}; m=1,..nl_i\}$, for each sublevel l_i. The ordering is generated by a Boltzmann machine, which represents the whole organization level, and measures the flow of knowledge $R_{im,i+1i}$ from node of sublevel l_i to node l_{i+1},

$$R_{im,i+1j} = -\tfrac{1}{2} w_{im,i+1i} q_{im} q_{ii+1j} \qquad (9.1)$$

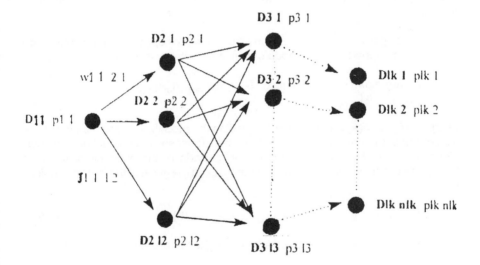

Fig. 9.4 Boltzmann Machine Representation for Product Scheduling

where the connection weight $w_{im,i+1j} > 0$

The transition probability is calculated to satisfy **Jaynes' maximum principle:**

$$p_{im,ii+1j} = \exp[-a_{im} -\tfrac{1}{2}w_{im,i+1j}\, q_{im}\, q_{ii+1j} \qquad\qquad \sum_{j=1}^{nll+1} p_{im,ii+1j} = 1 \qquad (9.2)$$

with a_{im} an appropriate normalization constant.

The negative entropy, in Shannon's sense, transferred from node d_{im} to d_{i+1j}, is

$$H_{im,i+1j} = -E\{\ln p_{im,ii+1j}\} = \beta_{im} + \tfrac{1}{2}w_{im,i+1j}\, q_{im}\, q_{ii+1j} \qquad (9.3)$$

where β_{im} is an appropriate normalization constant. $H_{im,i+1i}$, is an increasing function of probabilities and weights, and defines the order of concatenation of the nodes.

<u>Decision Making</u>, is associated with the Boltzmann Machine, and is obtained as follows:

• Starting at the node d_{im} the connections to nodes d_{i+1j} are searched until node j^*, corresponding to the node of minimum entropy is found.

$$j^* = \arg\max_j H_{im,i+1j} = \arg\max_j (\beta_{im} + \tfrac{1}{2}w_{im,i+1j}\, q_{im}\, q_{ii+1j})$$
$$H_{im,i+1j}^* = [H_{im,i+1j}]_{j=j^*}. \qquad (9.4)$$

<u>Learning (Feedback)</u> is obtained by upgrading the probabilities and weights of the successful path after each iteration. A double stochastic approximation recursive algorithm is used for the upgrading. The upgrade of the estimate of the cost function between nodes d_{im} and d_{i+1j}, namely $J_{im,i+1j}(t)$, is given at the (t+1)st iteration, by:

$$J_{im,i+1j}(t+1)= J_{im,i+1j}(t) + (t+1)^{-1}[J_{obs}(t_k+1) - J_{im,i+1j}(t)] \qquad (9.5)$$

where $J_{im,i+1j}(t)$ is the performance estimate, J_{obs} is the observed value and

$$p_{im}(t+1) = p_{im}(t) + (t+1)^{-1}[p - p_{im}(t)]$$
$$w_{dim,di+1j}(t+1) = w_{dim,di+1j}(t) + (t+1)^{-1}[w - w_{dim,di+1j}(t)] \qquad (9.6)$$

$$p, w = \Gamma_k(t_k+1) = \begin{cases} 1 & \text{if } J = \min \Sigma J \\ 0 & \text{otherwise} \end{cases} \qquad (9.7)$$

All these operations are performed by the Boltzmann Machine presented in Chapter 5, and shown in Fig. 9.5

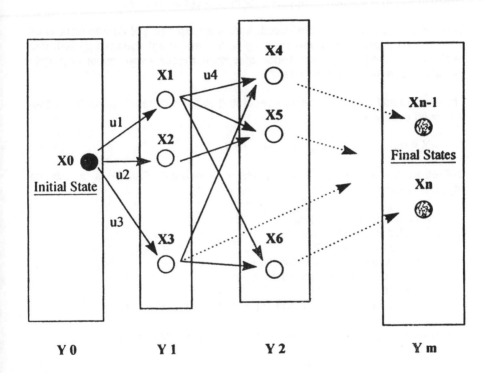

Fig. 9.5 A Finite State Machine Generator

9.3.2 Product Scheduling Architecture: The Coordination Level

The Coordination level (Varvatsoulakis et. Al. 1999) serves in this case as the interface between the Organization and the Execution levels. It maps the abstract tasks into the real world, without involving data packages left for the Execution level. Its **dispatcher** interprets and decomposes the commands received by the Organization level and distributes them to the appropriate coordinators which correspond to the stations of the factory. They in turn translate the control commands into operation instructions for the execution devices.

The class of coordinators, considered here, is an extension of deterministic finite state machine generators, representing a discrete event dynamic system that evolves according to the occurrence of a spontaneous event generating state transitions (Ramage, Wonham 1987).

The associated finite state space is generated by the n-dimensional state varibles X defined over a set SX.

$$X = [x_1,...x_n] \; ; \; x_i \in SX, \quad I=1,...n \quad n \in N \tag{9.8}$$

Definition 9.1 A Finite State Machine Generator (FSMG) is a sixtuple (X,U,f,g,X_0,X_f) where (Fig. 9.5)

X is the finite state space
U is the alphabet of events
Y is the output alphabet
$f:XxU \to X$ is the transition function
$g:XxU \to Y$ is the output function
$X_{ij} \in X$ is the initial state
$X_f \subseteq X$ is the set of final states representing completed tasks.

The dynamics of the FSMG is described by the difference equation:

$$X(k+1) = f(X(k), U(k))$$
$$Y(k+1) = g(X(k+1), U(k)); \tag{9.9}$$

where $X(k+1) \in X$, the state after the kth event; $U(k) \in U$, the kth event; and $Y(k+1) \in Y \subset 2^Y$ the set of all possible output symbols.

In terms of a formal regular language let:

$U^* \subset U$ denote the set of all finite strings including the empty string e, and the sequence at instance k, $U^*(k) = \{U(k),...U(0)\}$.

Letting f:XxU→X be extended to a function, the internal behavior of FSMG is described by a formal language, denoting the traces of events accepted by FSMG:

$$L_x(FSMG) \subseteq U^* := \{u^* \in U^* \mid f(u^*, X_0) \text{ is defined}\} \tag{9.10}$$

The Language L_{xf} of all finite traces representing the complete tasks generated by FSMG is defined as:

$$L_{xf}(FSMG) \subseteq L_x(FSMG) := \{ u^* \in L_x(FSMG) \mid f(u^*, X_0) \in X_f\} \tag{9.11}$$

In a similar way let Y^* of all finite strings over Y including the empty string e and $Y^*(k)=\{Y(k),...Y(0)\}$ containing the output symbols at time k. Letting h:$Y^* xX \to X$, the output

behavior of FSMG is described in terms of the formal regular language, denoting all finite traces of output symbols generated by FSMG.

$$L_Y(FSMG) \subseteq Y^* := \{y^* \in Y^* \mid y^* = h(u^*, X_0) \text{ is defined}, u^* \in U^*\} \tag{9.12}$$

The Language L_{Yf} representing the finite traces of output symbols complete tasks generated by FSMG is defined as:

$$L_{Yf}(FSMG) \subseteq L_Y(FSMG) := \{ y^* \in L_Y(FSMG) \mid h(y^*, X_0) \text{ is defined}, f(u^*, X_0) \in X_f\}$$
$$\tag{9.13}$$

A sequence of states $\{X(k),...X(0)\}$ is called a *path* of FSMG if there exists a sequence of event symbols $\{U(k),...U(0)\}$ such that $X(k+1)=f(X(k)U(k))$, k=1,...n-1. When X(k)= X(0) this path is called a **cycle**. If an FSMG has no cycles it is called *acyclic,* which is the case of our coordinators.

In the linguistic decision approach, task processes of a FSMG are the processes of translation

$$L_x(FSMG) \to L_y(FSMG) \quad \text{and}$$
$$L_{xf}(FSMG) \to L_{yf}(FSMG) \tag{9.14}$$

The formal languages $L_x(FSMG)$ and $L_{xf}(FSMG)$ define the control languages for a level, which through FSMG generate the set of control commands for the lower level.

In standard control terminology FSMG plays the role of the open loop plant with a certain physical behavior. The control mechanism available to the supervisor is the ability to block the occurrence of some events, in order that the whole system meets certain specifications

Definition 9.2. An FSMCG = (X,U,f,g,h,X_0,X_f) is a **finite state machine control generator** where

> X is the finite state space
> U is the alphabet of events
> Y is the output alphabet
> $f:XxU{\rightarrow}X$ is the transition function
> $h:XxU{\rightarrow}Y$ is the output function
> $g::XxU{\rightarrow}\{0,1\}$ is the control function
> $X_0{\in}X$ is the initial state
> $X_f{\subseteq}X$ is the set of final states representing completed tasks.

The behavior of the FSMCG is described by:

$$\begin{aligned}
X(k+1) &= f(X(k),\ U(k)) & &\text{if } g(X(k),\ U(k)) = 1 \\
X(k+1) &= X(k) & &\text{if } g(X(k),\ U(k)) = 1 \\
Y(k+1) &= h(X(k+1),\ U(k)); & &\text{if } g(X(k),\ U(k)) = 1 \\
Y(k+1) &= Y(k) & &\text{if } g(X(k),\ U(k)) = 0
\end{aligned} \tag{9.15}$$

Control of an FSMCG consists of switching the control function $g(X(k),U(k))$ according to defined conditions. If $g(X(k),U(k))=0$ the system is blocked and no state transition is allowed . This way every new event symbol $U(k)$ is rejected and not recorded in the event string. This controller is **static**.

Letting $g: X^* x U^* {\rightarrow} \{0,1\}$, the extended control function $g(X^*,U^*)$ depends on full state and event string observation. This controller is **dynamic**.

9.3.3 Product Scheduling Architecture: The Execution Level

The Execution level constitutes of the workstations corresponding one-to-one to the appropriate coordinators. Their effort is measured with a cost function interpreted as entropy, and fed back to the coordinators to make the selection of the proper path in product scheduling . They are part of the factory's hardware.

9.4. A PARADIGM OF AUTOMATED PRODUCTION SCHEDULING:.

A model of the production scheduling process is presented, for a particular class of problems, by the assembly of a gear box, from four different parts represented by the small square boxes in Fig. 9.6. It is based on the previously presented hierarchically intelligent product scheduling procedure.

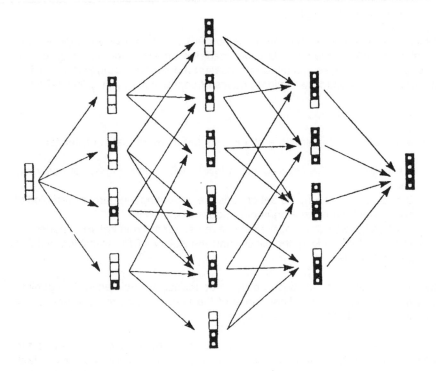

Fig. 9.6 Alternative Assembly Schedules

9.4.1 The Organization Level Structure

The main function of the Organization level is to construct a set of complete strings of events, that represent various tasks, as it was previously stated. It deals with the four operations ;

- Task representation
- Task Planning
- Decision Making
- Learning (Feedback)

A Boltzmann machine, described in section 9.3.1, can provide completely the optimal task representation, planning, decision making and learning of the sequence of events based on uncertain measurements. According to the specified it is possible to dynamically eliminate nodes or connections by properly defining

p,and w .

More particularly, for the structure of Fig. 9.6, the following sets are defined:

- $L = \{l_i, l=1,...n\}$ the ordered set of levels, with n the number of assembly parts.
- $D = \{d_{i1},...d_{in}; l=0,...n\}$ the set of nodes representing the parts assembled at each level l_i, where the number of nodes is n!/(n-i)!i!.
- $U = \{fetch part l; l=1,...n\}$ the set of events, transferring between nodes.
- The set P of node probabilities, and the set W of the transition weights between nodes.

A Memory storage area to back up the probabilities associated with organizer during the intermediate and final phase of the learning process is a necessary device to be part of the Organization level.

A search maximize the entropy between nodes is performed to generate the optimum string of the assembly procedure. After each iteration the cost of task execution is fed back from the Coordination level to upgrade the probabilities and the transition weights.

9.4.2 The Coordination Level Structure

The Coordination level is composed of a fixed *dispatcher* and several flexible **coordinators**. The dispatcher , modeled by an FSMG, receives from the Organizer the sequences of events as a string of commands and translates them into control actions for the corresponding Coordinators as control actions (Fig.9.7) Each

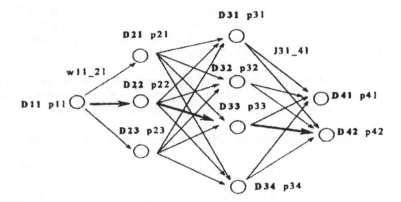

Fig. 9.7 The Manufacturing Network

Coordinator, modeled as a Finite State Machine Control Generator (FSMCG), uses the control actions to produce instructions transmitted to a **communication bus** for the initialization of other coordinators.

Let $SX = \{q_i; i=1,...n\}$ contain the parts of the assembly. If $m \le n$ is the number of parts already assembled the state of FSMG is the vector $X_m = \{x_i; l=1,...m; x_i \in SX\}$, representing the parts assembled and $X_n = \{x_i=0; l=m+1,...n; x_i \in SX\}$. The set of events $U=q \in SX$ and $X(k)=\{X_m(k),X_n(k)\}$, define the function

$$X(k+1) = f(X(k), U(k))$$

while $U(k) = \{x_1,...x_m,x_{m+1},0,...0\}$ for $x_{m+1}=q$. A scheduling procedure for the coordinators can be designed based on the output function of FSMG. Symbols from the dispatcher are read by each coordinator through the communication bus. They switch the control function $g(X(0),U(0))$ from 0 to 1to start the process; then it continues as long as $g(X(k),U(k))=1$ and blocks it when $g(X(k),U(k))=0$, expecting appropriate symbols from the other coordinators to switch back.

In this coordination structure, equivalent to the one proposed by Wang and Saridis (1990), the dispatcher has a dominant position, serving as the task control center and the only information communication chanel among the coordinators.

After each execution the performance update is evaluated with the formula:

$$J_i(k+1)= J_i(k) + (k+1)^{-1}[J_{obs}(k) - J_i(k)] \tag{9.16}$$

where $J_{obs}(k)$ is the measured performance at the kth iteration. The node probabilities are updated using the expression:

$$p_i(k+1) = p_i(k) + (k+1)^{-1}[p - p_i(k)]$$

$$p = \begin{cases} 1 & \text{if } J = \min J_i(k), i=1,...m \\ 0 & \text{otherwise} \end{cases} \tag{9.17}$$

9.4.3 The Execution Level Structure

The Execution level consists of groups of devices corresponding one-to-one with the appropriate coordinators. The subcosts for each device are measured as entropies, after each complete execution of the task. Then they are transferred BACK to the coordination level for computation of the overall measured cost and upgrading of the upper levels.

9.5 SIMULATION RESULTS FOR THE ASSEMBLY OF A MACHINE

This special case study involves the simulation of a multilevel decision making system based on the assembly of four different parts of a machine described in Fig.9.7. The black boxes in that figure represent assembled parts.

The Boltzmann machine for the a number of schedules. The maximum entropy schedule is transmitted to the Coordination level.

There are four nodes D and the total number of paths, representing alternate production plans, is 24. All weight and probability limits are equal to 1. The costs of all paths are observed and estimated during every iteration. The optimal schedule is represented with the bold lines in Fig. 9.7.

The minimum estimated cost in terms of number of iterations is given in Fig.9.8. The maximum number of iteration to converge to the optimal sequence of a defined path is about 250.

9.6. REMARKS

The two case studies presented in this Chapter, exemplify the use and validity of Hierarchically Intelligent Machines, under two different architectures. It also indicates the trend of the future of Automation.

More applications, whenever they are accepted by the industry, will completely establish the true need of such an approach. It should be reminded again, that the software structures are flexible and may be matched to the taste of the designer.

9.7 REFERENCES

The Competitive Edge: Research Priorities for U.S. Manufacturing, (1989) Report of the National Research Council on U.S. Manufacturing, National Academy Press.

Ramage P. J., and Wonham W. M., (1987), "Suprvisory Control of a class of discrete event processes" *SIAM Journal of Control Optimization*, Vol. 25, No. 1, pp.206-230

Saridis G. N. (1996)"Architectures for Intelligent Controls" in *Intelligent Control Systems*, Edited by Madan. M. Gutta and Naresh K. Sinha, Chapter 6, pp.127-148, IEEE Press.

Saridis G. N. (1979), "Toward the Realization of Intelligent Controls", *IEEE Proceedings*, 67, No. 8.

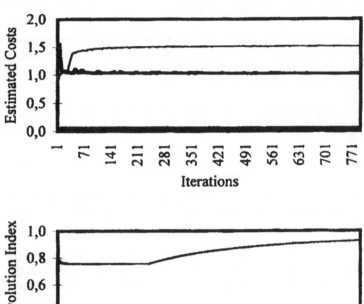

Fig. 9.8 Learning Curves for Assembly Scheduling

Saridis, G.N. and Graham, J.H. (1984), "Linguistic Decision Schemata for Intelligent Robots", *AUTOMATICA the IFAC Journal,* 20, No. 1, pp. 121-126, Jan.

Saridis, G.N. and Moed, M.C. (1988), "Analytic Formulation of Intelligent Machines as Neural Nets", *Symposium on Intelligent Control*, Washington, D.C., August. Wang, F., Saridis, G.N.(1990) "A Coordination Theory for Intelligent Machines" *AUTOMATICA the IFAC Journal,* 35, No. 5, pp. 833-844, Sept.

Valavanis, K.P., Saridis, G.N.(1992), *Intelligent Robotic System Theory: Design and Applications*, Kluwer Academic Publishers, Boston MA.

Varvatsoulakis M., Saridis G. N., and Paraskevopoulos P., (1998)"A Model for the Organization Level of Intelligent Machines" *Proceedings of 1998 International Conference on Robotics and Automation*, Leuven Belgium May 15.

M. N. Varvatsoulakis, G. N. Saridis, P. N. Paraskevopoulos, (2000), "Intelligent Organization for Flexible Manufacturing" *IEEE Transactions on Robotics and Automation* Vol.16, No. 2, pp. 180-189, April.

Chapter 10.

CONCLUSIONS

10.1 DISTRIBUTED INTELLIGENT MACHINES AND COOPERATING ROBOTS

The Hierarchically Intelligent Machines, described in this volume, have been designed as single operators in structured or unstructured environments. However, many of them, may be asked to work as a team to perform complex task. In that case they should be coordinated at a higher level in order to produce satisfactory results (Boetcher and Levis 1983). Example is the autonomous construction of the Space Station in space, by a team of Hierarchically Intelligent Robots.

Team coordination is a rather complex task, requiring command interpretation and distribution, and assignment and sequencing of activities of the individual robots. This is a problem that will not be solve d in the present work. However some suggestions will be made in order to guide future research.

Assuming that the machines under consideration will not be antagonistic to each other, two possible configurations for the dissemination of commands necessary to execute the complex task.

- **Star Connection**
- **Ring Connection**

The first one agrees more with the hierarchical architecture of the Hierarchically Intelligent Machines that were presented in the previous Chapters. The command enters a central node of the system, where it is processed, and is distributed radially to the peripherally distributed machines for execution.

The second one represents a more democratic point of view, where decisions about the distribution of executions is done by majority vote.

Both of methods have their pros and cons and should be selected according to the needs of the specific task.

10.2 FUTURE RESEARCH

The architecture described in this paper does not differ substantially from the architecture originally proposed by Saridis (1983). The details have been more elaborated and more efficient internal structures have been used. (Fig.10.1)The main contribution though is that this system has been successfully implemented and that the resulting structure is extremely efficient, effective, versatile, capable for

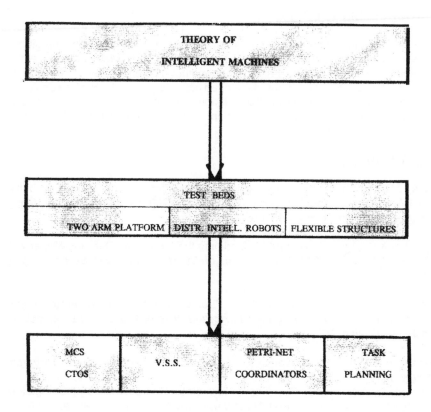

Fig. 10.1 A View of the New Intelligent Architecture for Robotic Assembly

remote operation as compared to other proposed architectures.

Future work should be directed towards the implementation of various algorithms to suit the hierarchically architecture for different tasks meant to replace human efforts in hazardous, unfamiliar, and remote environments. Special emphasis should be given to distributed and cooperating architectures of Hierarchically Intelligent Machines which are getting more and more popular.

10.3 CONCLUDING REMARKS

However, new applications of Hierarchically Intelligent Machines, are emerging in areas of soft sciences Biotechnology, Economics, Ecology, Manufacturing, etc as was pointed out in previous Chapters. The methodologies are flexible enough, to accommodate various algorithms, developed according to the taste of the designer, assigned to the Hierarchically Intelligent Machine structure, for these sciences and technologies. More work in this direction, will prove the validity of the effort generated in this book.

10.4. REFERENCES

Albus, J.S. (1975), "A New Approach to Manipulation Control: The Cerebellar Model Articulation Controller", *Transactions of ASME, J. Dynamics Systems, Measurement and Control,* 97, pp. 220-227.

Boettcher K. L., Levis A. H., (1983),"Modeling the Interacting Decision-Maker with Bounded Rationality", *IEEE Transactions on System Man and Cybernetics, Vol. SMC-12*, 3, pp.

Hinton, G.E. and Sejnowski, T.J. (1986), "Learning and Relearning in Boltzmann Machines", pp. 282-317, in *Parallel Distributed Processing*, eds. D.E. Rumelhart and J.L. McClellan, MIT Press.

Jaynes, E.T. (1957), "Information Theory and Statistical Mechanics", *Physical Review*, pp. 106, 4.
Kolmogorov, A.N. (1956), "On Some Asymptotic Characteristics of Completely Bounded Metric Systems", *Dokl Akad Nauk, SSSR*, 108, No. 3, pp. 385-389.

McInroy J.E., Saridis G.N.,(1991), "Reliability Based Control and Sensing Design for Intelligent Machines", in *Reliability Analysis* ed. J.H. Graham, Elsevier North Holland, N.Y.

Meystel, A. (1986), "Cognitive Controller for Autonomous Systems", *IEEE Workshop on Intelligent Control 1985*, p. 222, RPI, Troy, New York.

Mittman M.,(1990), "TOKENPASSER: A Petri Net Specification Tool" *Master's Thesis, Rensselaer Poly. Inst., Dec.*

Moed, M.C. and Saridis, G.N. (1990), "A Boltzmann Machine for the Organization of Intelligent Machines", *IEEE Trans. on Systems Man and Cybernetics, 20*, No. 5, Sept.

Nilsson, N.J. (1969), "A Mobile Automaton: An Application of Artificial Intelligence Techniques", *Proc. Int. Joint Conf. on AI*, Washington, D.C.

Peterson, J.L. (1977), "Petri-Nets", *Computing Survey, 9*, No. 3, pp. 223-252, September.

Saridis, G.N. (1977), *Self-Organizing Controls of Stochastic Systems*, Marcel Dekker, New York, New York.

Saridis, G.N. (1979), "Toward the Realization of Intelligent Controls", *IEEE Proceedings, 67*, No. 8.

Saridis, G. N. (1983), "Intelligent Robotic Control", *IEEE Trans. on AC, 28*, 4, pp. 547-557, April.

Saridis, G.N. (1985), "Control Performance as an Entropy", *Control Theory and Advanced Technology, 1*, 2, pp. 125-138, Aug.

Saridis, G.N. (1985), "Foundations of Intelligent Controls", *Proceedings of IEEE Workshop on Intelligent Controls*, p. 23, RPI, Troy, New York.

Saridis, G.N. (1988), "Entropy Formulation for Optimal and Adaptive Control", *IEEE Transactions on AC, 33*, No. 8, pp. 713-721, Aug.

Saridis, G.N. (1989), "Analytic Formulation of the IPDI for Intelligent Machines", *AUTOMATICA the IFAC Journal, 25*, No. 3, pp. 461-467.

Saridis, G.N. and Graham, J.H. (1984), "Linguistic Decision Schemata for Intelligent Robots", *AUTOMATICA the IFAC Journal, 20*, No.1, pp. 121-126, Jan.

Saridis, G.N. and Moed, M.C. (1988), "Analytic Formulation of Intelligent Machines as Neural Nets", *Symposium on Intelligent Control*, Washington, D.C., August.

Saridis, G.N. and Stephanou, H.E. (1977), "A Hierarchical Approach to the Control of a Prosthetic Arm", *IEEE Trans. on SMC, 7*, No. 6, pp. 407-420, June.

Saridis, G.N. and Valavanis, K.P. (1988), "Analytical Design of Intelligent

Machines", *AUTOMATICA the IFAC Journal*, *24*, No. 2, pp. 123-133, March.

Shannon, C. and Weaver, W. (1963), *The Mathematical Theory of Communications*, Illini Books.

Varvatsoulakis M., Saridis G. N., and Paraskevopoulos P., (1998)"A Model for the Organization Level of Intelligent Machines" *Proceedings of 1998 International Conference on Robotics and Automation*, Leuven Belgium May 15.

Varvatsoulakis M., Saridis G. N., and Paraskevopoulos P., (1999) "Intelligent Structures for Automated Assembly Scheduling" Submitted to *IEEE Transactions on Control Systems Technology.*

Wang, F., Kyriakopoulos, K., Tsolkas T., Saridis, G.N., (1990) "A Petri-Net Coordination Model of Intelligent Mobile robots" *CIRSSE Technical Report #50*, Jan.

Wang, F. and Saridis, G.N. (1988), "A Model for Coordination of Intelligent Machines Using Petri Nets", *Symposium on Intelligent Control*, Washington, D.C., August.

Wang, F., Saridis, G.N. (1990) "A Coordination Theory for Intelligent Machines" *AUTOMATICA the IFAC Journal*, *35*, No. 5, pp. 833-844, Sept.

Zames, G. (1979), "On the Metric Complexity of Casual Linear Systems, ϵ-entropy and ϵ-dimension for Continuous Time", *IEEE Trans. Automatic Control*, *24*, No. 2, pp. 220-230, April.

INDEX